CW00456482

REAL VARIABLES WITH BASIC METRIC SPACE TOPOLOGY

Robert B. Ash
Department of Mathematics
University of Illinois at Urbana-Champaign

DOVER PUBLICATIONS, INC.
Mineola, New York

Copyright

Copyright © 2007 by Robert B. Ash.
All rights reserved.

Bibliographical Note

This Dover edition, first published in 2009, is the first publication in book form of the revised edition of *Real Variables with Basic Metric Space Topology* by Robert B. Ash, originally published in 1993 by IEEE Press, New York. This edition is also available from the following web site: http://www.math.vivc.edu/~r-ash/

Library of Congress Cataloging-in-Publication Data

Ash, Robert B.
Real variables with basic metric space topology / Robert B. Ash. — Dover ed.
p. cm.
Originally published: New York : IEEE Press, 1993.
Includes index.
ISBN-13: 978-0-486-47220-1
ISBN-10: 0-486-47220-5
1. Functions of real variables. 2. Metric spaces. I. Title.

QA331.5.A78 2009
515'.8—dc22

2009000501

Manufactured in the United States of America
Dover Publications, Inc., 31 East 2nd Street, Mineola, N.Y. 11501

CONTENTS

PREFACE

This is a text for a first course in real variables. The subject matter is fundamental for more advanced mathematical work, specifically in the areas of complex variables, measure theory, differential equations, functional analysis, and probability. In addition, many students of engineering, physics, and economics find that they need to know real analysis in order to cope with the professional literature in their fields. Standard mathematical writing, with its emphasis on formalism and abstraction, tends to create barriers to learning and focus on minor technical details at the expense of intuition. On the other hand, a certain amount of abstraction is unavoidable if one is to give a sound and coherent presentation. This book attempts to strike a balance that will reach the widest audience possible without sacrificing precision. (My original training was in electrical engineering, and I later became a mathematician. I seem to be able to reach students whose prior exposure to abstract reasoning is limited.)

The most important skill a student must develop in order to learn any area of mathematics is the ability to think intuitively, and a text should encourage this process–not hinder it. I find it useful to consider concrete examples that have all the features of the general case under consideration, to draw diagrams whenever appropriate, and to give geometric or physical interpretation of results. I rely especially on one of the most useful of all learning devices: the inclusion of detailed solutions to exercises. Solutions to problems are commonplace in elementary texts but quite rare (although equally valuable) at the upper division undergraduate and graduate level. This feature makes the book suitable for independent study, and further widens the audience.

I have normally covered the first seven chapters in a one-semester course. The subject matter includes metric spaces, Euclidean spaces and their basic topological properties, sequences and series of real numbers, continuous functions, differentiation, Riemann–Stieltjes integration, and uniform convergence and applications. Chapters 8 and 9 contain additional topological results, and various sections might be assigned in connection with special student projects or serve as a nice introduction to a full course in general topology.

The subject matter covered in basic courses in real variables is standard, almost canonical, but there is still room for innovation in the presentation. I have taken Cantor's Nested Set Property, rather than Cauchy completeness or least upper bound completeness, as the basic topological axiom for the real numbers. This allows a significant simplification of the proofs of the basic topological properties of R^p.

1

INTRODUCTION

1.1 BASIC TERMINOLOGY

In a course in real analysis, the normal procedure is to begin with a definition of the real numbers, either by means of a set of axioms or by a constructive procedure which starts with the "God–given" set of positive integers. The set of all integers is constructed, and from this the rational numbers are obtained, and finally the reals. A discussion of this type is part of the area of logic and foundations rather than real analysis, and we will postpone it until much later. For now we'll take the point of view that we know what the real numbers are: A *real number* is an integer plus an infinite decimal, for example, $65.7204\ldots$

If the decimal ends in all nines, we have two representations of the same real number, for example,

$$5.237999\cdots = 5.2380000\ldots.$$

We will often talk about sets of real numbers, and therefore a modest amount of set–theoretic terminology is necessary before we can get anywhere. You have probably seen most of this in another course, so we will proceed rather quickly.

1.1.1 Definitions and Comments

The *union* of two sets A and B, denoted by $A \cup B$, is the set of points belonging to *either A or B* (*or both*; from now on, the word "or" always has the so-called *inclusive* connotation "or both" unless otherwise specified).

The *intersection* of two sets A and B, denoted by $A \cap B$, is the set of points belonging to *both A* and B.

The *complement* of a set A, denoted by A^c, is the set of points *not* belonging to A. (Generally, we will be working in a fixed space Ω (sometimes called the *universe*), for example, the set of real numbers or perhaps the set of pairs of real numbers, that is, the Euclidean plane. All sets under discussion will consist of various points of Ω, and thus A^c is the set of points of Ω that do not belong to A).

Unions, intersections, and complements may be represented by pictures called *Venn diagrams* that are probably familiar to many readers; see Fig. 1.1.1.

Unions and intersections may be defined for more than two sets, in fact for an arbitrary collection of sets.

The *union* of sets $A_1, A_2, \ldots$, denoted by $A_1 \cup A_2 \cup \ldots$ or by $\bigcup_i A_i$, is the set of points belonging to *at least one* of the A_1; the *intersection* of $A_1, A_2, \ldots$, denoted by $A_1 \cap A_2 \cap \ldots$ or by $\bigcap_i A_i$, is the set of points belonging to *all* the A_i. The union of sets $A_1, \ldots, A_n$ is often written as $\bigcup_{i=1}^n A_i$, and the union of an infinite sequence $A_1, A_2, \ldots$ is denoted by $\bigcup_{i=1}^\infty A_i$, with similar notation for intersection.

There are a few identities involving unions, intersections, and complements that come up occasionally. For example, the *distributive law* holds: for arbitrary sets A, B, C,

$$A \cap (B \cup C) = (A \cap B) \cup (A \cap C) \tag{1}$$

(the word "distributive" is used because in this formula, at least, intersection behaves like ordinary multiplication and union like addition).

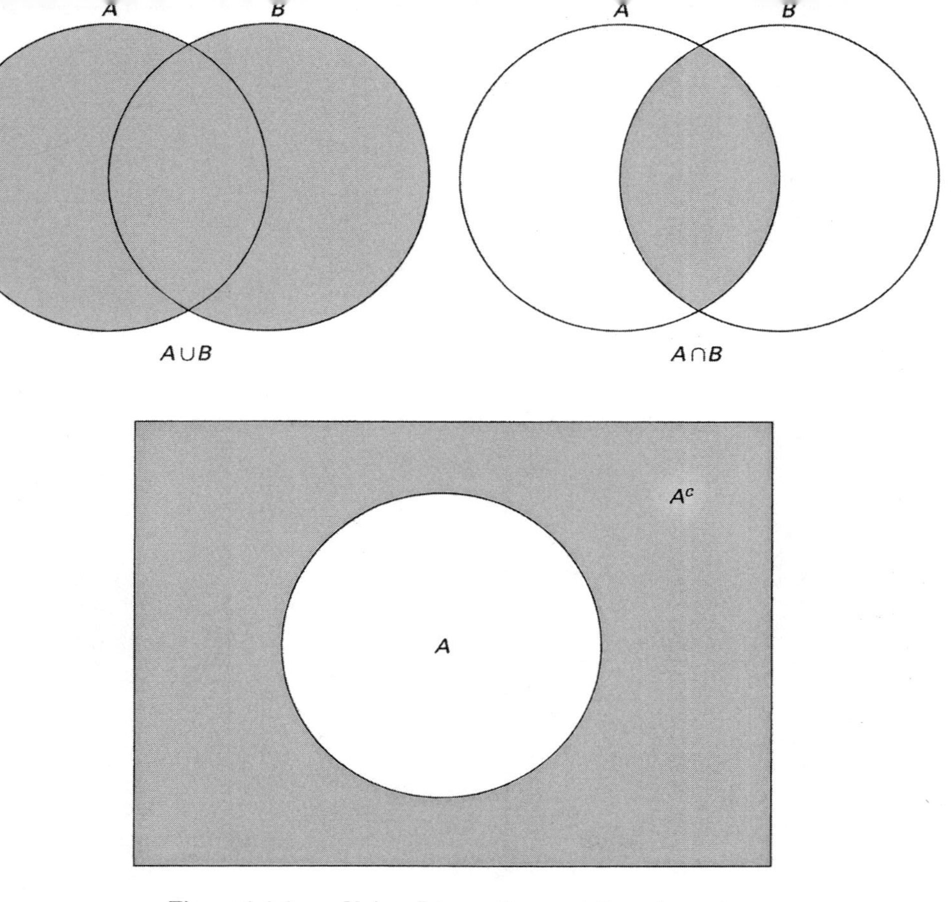

Figure 1.1.1 Union, Intersection, and Complement

The formula may be verified by drawing a Venn diagram (Fig. 1.1.2) or by showing that the sets on the left and right sides of the equality have the same members, as follows. (The symbol $\in$ means "belongs to".) If $x \in A \cap (B \cup C)$, then $x \in A$ and $x \in B \cup C$, so that $x \in B$ or $x \in C$.

Case 1. $x \in A$ and $x \in B$; then $x \in A \cap B$; hence $x \in (A \cap B) \cup (A \cap C)$.

Case 2. $x \in A$ and $x \in C$; then $x \in A \cap C$; hence $x \in (A \cap B) \cup (A \cap C)$.

Caution. So far we have shown only that the set on the left is a *subset* of the set on the right. In general, the set D is said to be a subset of the set E if every point of D also belongs to E (see Fig. 1.1.3). We use the notation $D \subseteq E$; sometimes we say that E *contains* D or the D is *contained in* E. In order to show that $D = E$ we must also show that every point of E belongs to D. If $D \subseteq E$ but $D \neq E$, we say that D is a *proper subset* of E, sometimes written $D \subset E$. Note that according to the definition, a set is a subset of itself.

To return to the proof of the distributive law, suppose $x \in (A \cap B) \cup (A \cap C)$. Then $x \in A \cap B$ or $x \in A \cap C$; thus, we know definitely that $x \in A$, and also $x \in B$ or $x \in C$, as desired.

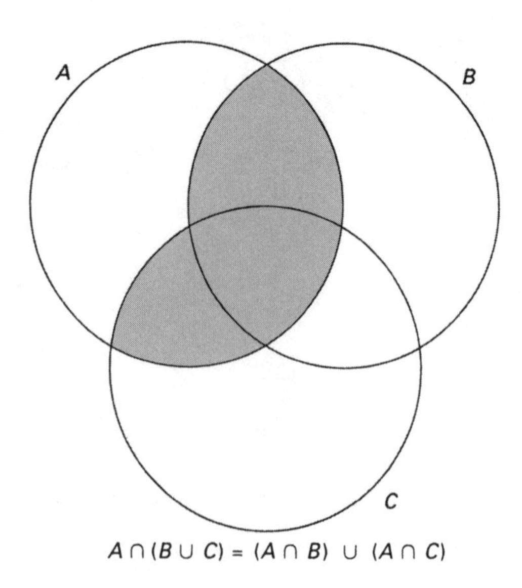

Figure 1.1.2 Distributive Law

Figure 1.1.3 Subset Relation

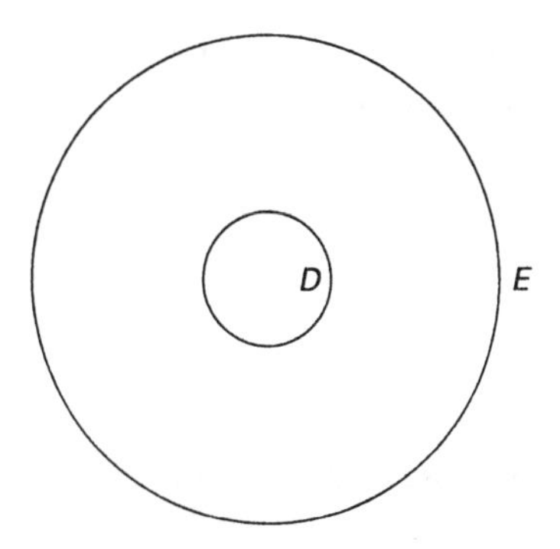

The following identities, called the *De Morgan laws*, are often useful:

(2)
$$(A \cap B)^c = A^c \cup B^c,$$
$$(A \cup B)^c = A^c \cap B^c.$$

As above, the identities can be verified by a Venn diagram or a formal argument using the definitions of union, intersection, and complement. In fact the De Morgan laws can be extended to an *arbitrary* collection of sets, as follows:

(3)
$$\Big(\bigcap_i A_i\Big)^c = \bigcup_i A_i^c,$$
$$\Big(\bigcup_i A_i\Big)^c = \bigcap_i A_i^c.$$

The Venn diagram approach is not useful with such a large collection of sets, and the formal method must be used (Problem 1).

To complete the necessary set–theoretic terminology, the *empty set* (set with no members) will be denoted by $\emptyset$. If A is any set, then $\emptyset \subseteq A$. (If you produce an element $x \in \emptyset$, I will be delighted to show that x belongs to A.) We will look at this idea in more detail in Section 1.4. The sets A_i are said to be *disjoint* or *mutually exclusive* if there is no overlap between *any pair* of sets; that is,

$$A_i \cap A_j = \emptyset \qquad \text{whenever} \quad i \neq j.$$

As an example, let $A = \{1,2,3\}$, $B = \{1,2\}$, $C = \{3,5\}$ (the notation is standard: a set is described by listing its members, so that, for example, A is the set consisting of the numbers 1, 2, and 3). In this case, A, B, C are *not disjoint*, although $A \cap B \cap C = \emptyset$. For disjointness we must have $A \cap B = A \cap C = B \cap C = \emptyset$, and here we have $A \cap B = \{1,2\} \neq \emptyset$, $A \cap C = \{3\} \neq \emptyset$. Since $B \cap C = \emptyset$, we may say that B and C are disjoint.

The *set–theoretic difference* between A and B is defined by

$$\begin{aligned} A \backslash B &= \{x \in \Omega\colon x \in A, x \notin B\} \\ &= A \cap B^c. \end{aligned}$$

Problems for Section 1.1

1. Prove the De Morgan laws for an arbitrary collection of sets (Eq. (3)).
2. Prove that union distributes over intersection; i.e.,

$$A \cup (B \cap C) = (A \cup B) \cap (A \cup C).$$

3. Show that the union of three arbitrary sets can be written as a disjoint union in the following way:

$$A \cup B \cup C = A \cup (A^c \cap B) \cup (A^c \cap B^c \cap C).$$

4. Continuing Problem 3, if $A_1, A_2, \ldots$ are arbitrary sets, show that

$$\bigcup_{n=1}^{\infty} A_n = A_1 \cup (A_1^c \cap A_2) \cup \cdots \cup (A_1^c \cap \cdots \cap A_{n-1}^c \cap A_n) \cup \ldots.$$

5. If A, B, C, D are arbitrary sets, express the set of elements belonging to at least two of the sets A, B, C, D, using unions and intersections of A, B, C, D.
6. Repeat Problem 5 with *at least two* replaced by *exactly two* and use *complements* as well as unions and intersections.

1.2 FINITE AND INFINITE SETS; COUNTABLY INFINITE AND UNCOUNTABLY INFINITE SETS

Sometimes we need to know something about the size of a set, in particular, whether it is finite. If a set is infinite, it may be useful to know if it can somehow be counted or if it is uncountable. The definitions are as follows.

A *finite set* is one that can be put in one–to–one correspondence with $\{1, 2, \ldots, n\}$ for some positive integer n (by convention, the empty set is regarded as finite). An *infinite set* is a set that is not finite. A *countably infinite set* is one that can be put in one–to–one correspondence with the entire set of positive integers. This means simply that the points can be labeled 1, 2, 3, ... A set is *uncountably infinite* if it is infinite but not countably infinite. It is convenient to call a set *countable* if it is either finite or countably infinite; thus, *uncountable* is synonymous with uncountably infinite.

Possibly you have seen the classic arguments that the set of rational numbers is countably infinite, but the set of all real numbers is uncountably infinite. To count the positive rationals, we devise an explicit scheme (see Fig. 1.2.1). The procedure amounts to counting the points of an infinite rectangular array, and is slightly inefficient because each rational number appears infinitely often: for example,

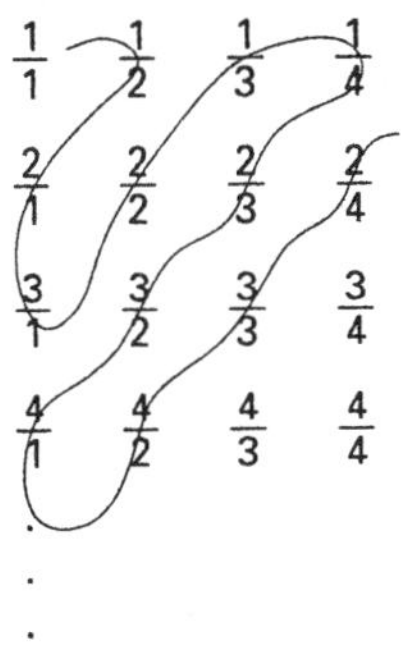

$$\begin{array}{ccccc} \frac{1}{1} & \frac{1}{2} & \frac{1}{3} & \frac{1}{4} & \cdots \\ \frac{2}{1} & \frac{2}{2} & \frac{2}{3} & \frac{2}{4} & \cdots \\ \frac{3}{1} & \frac{3}{2} & \frac{3}{3} & \frac{3}{4} & \cdots \\ \frac{4}{1} & \frac{4}{2} & \frac{4}{3} & \frac{4}{4} & \cdots \\ \vdots & & & & \end{array}$$

$r_1 = 1$, $r_2 = \frac{1}{2}$, $r_3 = 2$, $r_4 = 3$, $r_5 = \frac{1}{3}$, $r_6 = \frac{1}{4}$, $r_7 = \frac{2}{3}$, $r_8 = \frac{3}{2}$, $r_9 = 4$, $r_{10} = 5$, $r_{11} = \frac{1}{5}$, ...

Figure 1.2.1 Counting the Rational Numbers

$1/2 = 2/4 = 3/6$, and so on. After the first appearance of a number (say $r_2 = 1/2$ in Fig. 1.2.1), all other appearances are skipped in making the count. To show that the reals are uncountable, we use the *Cantor diagonal process*. (This idea will occur several times later on.) Suppose we were able to count the real numbers between 0 and 1; list the numbers, in decimal form, as follows:

$$\begin{aligned} r_1 &= .a_{11}a_{12}a_{13}\cdots, \\ r_2 &= .a_{21}a_{22}a_{23}\cdots, \\ r_3 &= .a_{31}a_{32}a_{33}\cdots. \end{aligned} \tag{4}$$

Then form the real number $r = .b_1b_2b_3\ldots$, where $b_1 \neq a_{11}$, $b_2 \neq a_{22}$, $b_3 \neq a_{33}, \ldots$. To avoid the ambiguity caused by expansions ending in all nines or all zeros, we can, if we like, take $1 \leq b_n \leq 8$ for all n. Then r is a real number between 0 and 1, but cannot appear on the list.

There is an extensive theory of infinite sets, but we will be content with a few basic results.

1.2.1 THEOREM. *There are 2^n subsets of $\{1, 2, \ldots, n\}$.*

Proof. If $S \subseteq \{1, 2, \ldots, n\}$, then $1 \in S$ or $1 \notin S$ (*two* choices), $\ldots$, $n \in S$ or $n \notin S$ (*two* choices). The total number of subsets is the same as the total number of choices, namely, $2 \times 2 \times \cdots \times 2 = 2^n$. ∎

1.2.2 THEOREM. *There are uncountably many subsets of the positive integers.*

Proof. Make a correspondence between subsets of the positive integers and binary representations of real numbers between 0 and 1, as follows:

$$.10010110\ldots \quad \text{means} \quad 1 \in S, 2 \notin S, 3 \notin S, 4 \in S, 5 \notin S, 6 \in S, 7 \in S, 8 \notin S, \quad \text{etc.}$$

Since there are uncountably many reals between 0 and 1, there are uncountably many subsets of $\{1, 2, \ldots\}$.[1] ∎

[1]Note that the real number $.01000\cdots = .0011111\ldots$ actually yields *two* subsets, $\{2\}$ and $\{3, 4, 5, 6, 7, \ldots\}$. Since this *increases* the number of subsets relative to the reals, the conclusion of 1.2.2 is undisturbed.

1.2.3 THEOREM. *A countable union of countable sets is countable. In other words, if for each $n = 1, 2, \ldots, A_n$ is countable, then $\bigcup_{n=1}^{\infty} A_n$ is countable.*

Proof. List the members of the A_i as follows:

$$\begin{array}{llll} A_1: & a_{11} & a_{12} & \cdots \\ A_2: & a_{21} & a_{22} & \cdots \\ A_3: & a_{31} & a_{32} & \cdots \\ \vdots & & & \end{array}$$

Then count $\bigcup_{n=1}^{\infty} A_n$ by the same procedure we used to count the rationals. ■

Problems for Section 1.2

1. Give an alternative proof of Theorem 1.2.2, as follows. If $S_1, S_2, \ldots$ is a list of subsets of $\{1, 2, \ldots\}$, construct a subset S that cannot possibly be on the list.
2. Show that there are only countably many finite subsets of the positive integers.
3. Verify informally that the mapping

$$(x,y) \to \frac{1}{2}[(x+y)(x+y+1)+2x)]$$

is a one-to-one correspondence between ordered pairs (x,y) of nonnegative integers and nonnegative integers (see diagram).

⓪ (0, 0)	② (1, 0)	⑤ (2, 0)	⑨ (3, 0)	
① (0, 1)	④ (1, 1)	⑧ (2, 1)	. . .	
③ (0, 2)	⑦ (1, 2)	. . .		
⑥ (0, 3)				

4. The method of Fig. 1.2.1 shows that the positive rationals are countably infinite. How would you modify the procedure so as to count all the rationals?

5. Suppose that the rational numbers between 0 and 1 are listed as in (4). We then pick a rational $r = .r_1 r_2 r_3 \dots$ with $r_n \neq a_{nn}$, $n = 1, 2, \dots$. Why doesn't this show that the rationals are uncountable?
6. Show that it is impossible to list the rational numbers in increasing order.
7. Show that for any positive integer n the set of all $(x_1, \dots, x_n)$, where the x_i are rational, is countable.

1.3 DISTANCE AND CONVERGENCE

One of the basic ideas of analysis is that of *convergence*; a sequence of numbers x_n converges to a number x if, as n gets very large, x_n gets very close to x; in other words, the *distance* between x_n and x gets very small. The key concept is that of distance; as long as we have a distance function, we can talk about convergence. You are familiar with several distance functions. If x and y are points on the real line, the distance between them is $|x - y|$; if $x = (x_1, x_2)$ and $y = (y_1, y_2)$ are points in the plane, the Euclidean distance between them is $[(x_1 - y_1)^2 + (x_2 - y_2)^2]^{1/2}$. What properties must a distance function satisfy in order that we can talk about convergence sensibly? It turns out that only a few are needed, as follows.

1.3.1 Definitions and Comments

A *metric* or *distance function* on a set Ω is an assignment, to each pair of points (x, y), $x, y \in \Omega$, of a nonnegative real number $d(x, y)$, such that for all $x, y, z \in \Omega$ we have

$$(a) \qquad d(x,x) = 0; \qquad d(x,y) > 0 \quad \text{if} \quad x \neq y,$$

$$(b) \qquad d(x,y) = d(y,x),$$

$$(c) \qquad d(x,z) \leq d(x,y) + d(y,z).$$

Statement (c) is called the *triangle inequality*; if x, y, and z are vertices of a triangle in the plane, (c) says that the length of one side of the triangle cannot exceed the sum of the lengths of the other two sides.

A set Ω on which a distance function is defined is called a *metric space*. Our basic metric spaces will be the set of real numbers, to be denoted from now on by R, and *Euclidean p–space* R^p, the set of all p–tuples $(x_1,\dots,x_p)$ of real numbers. The metric on R^p is given by

$$d[(x_1,\dots,x_p),\quad (y_1,\dots,y_p)] = \left[\sum_{j=1}^{p}(x_j - y_j)^2\right]^{1/2}.$$

When $p = 1$, we have $R^p = R$, $d(x,y) = |x - y|$. We know because of our familiarity with elementary geometry that d is a metric when $p \leq 3$, but this must be *proved* when $p > 3$. However, it's probably best to wait until we have more experience before doing this, so let's accept the result for now. [See Section 3.2, Problem 5 for a proof.]

On any metric space there is a natural notion of *convergence*: if x_1, $x_2,\dots$ is a sequence of points in Ω, we say that the sequence converges to x (notation $x_n \to x$) if $d(x_n,x) \to 0$ as $n \to \infty$; in other words,

$$\lim_{n\to\infty} d(x_n,x) = 0.$$

The definition of limit might have been baffling when you first encountered it in calculus, but let's look at it again. The statement "$\lim_{n\to\infty} d(x_n,x) = 0$" means that given $\epsilon > 0$ there is a positive integer N such that if $n \geq N$ then $d(x_n,x) < \epsilon$. For example, suppose $\epsilon = 10^{-6}$. Then if we go far enough out in the sequence (it might turn out, say, that we have to go beyond $N = 10^8$), then all x_n from that point on (that is, $n \geq 10^8$) are within distance 10^{-6} of the point x. Thus after a certain point, all the x_n's are very close to x.

The distance function and the convergence concept allow us to study the structure of various sets in a metric space. For example, in R^2 the set of points whose distance from the origin is less than r is the interior of the circle of radius r and center at $(0,0)$. In general, an *open ball* in a metric space Ω is a set of the form

$$B_r(x) = \{y \in \Omega : d(x,y) < r\} \quad \text{where} \quad x \in \Omega \quad \text{and} \quad r > 0.$$

(Again, the set–theoretic notation is standard; $B_r(x)$ is the set of points y in Ω whose distance from x is less than r. When it is clear what space we are working in, we may omit "$\in \Omega$.") A *closed ball* is a set of the form

$$C_r(x) = \{y \in \Omega : d(x,y) \leq r\}.$$

Let E be a subset of the metric space Ω; E is said to be an *open set* if for every $x \in E$ there is an open ball $B_r(x)$ that is entirely contained in E; that is, $B_r(x) \subseteq E$.

For example, in R^2, $E = \{(x,y) : 0 < x < 1,\ y > 2\}$ is the interior of an infinite rectangular strip and is therefore open. Intuitively, an open set is one that does not contain any of its boundary points.

Now consider $E_1 = \{(x,y) : 0 < x < 1, y = 0\}$; E_1 is not open because an open ball *in* R^2 with center at $(x,0)$ cannot lie entirely within E_1 (Fig. 1.3.1). Note, however, that $A = \{x \in R : 0 < x < 1\}$ *is* open, because in R an open ball is just an open interval: $B_r(x) = \{y \in R : x - r < y < x + r\}$ (see Fig. 1.3.2). Thus, in talking about open sets, be sure you know which space you are working in. This applies also to closed sets, to be discussed next. Before going on, let's introduce the standard notation for open, closed, and semiclosed intervals of R:

$$\begin{aligned}
(a,b) &= \{x \in R : a < x < b\},\\
(a,b] &= \{x \in R : a < x \leq b\},\\
[a,b) &= \{x \in R : a \leq x < b\},\\
[a,b] &= \{x \in R : a \leq x \leq b\},\\
(a,\infty) &= \{x \in R : x > a\},
\end{aligned}$$

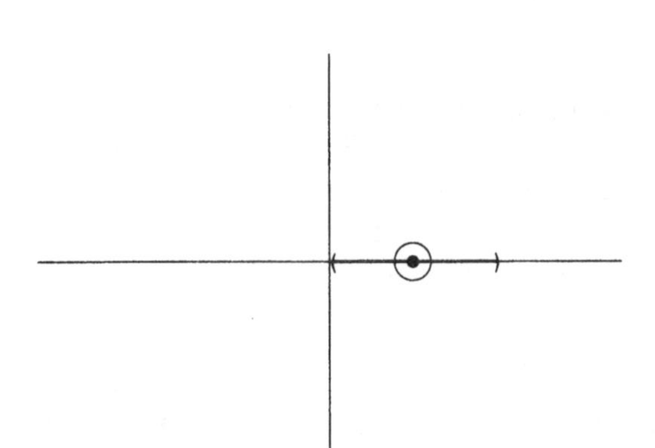

Figure 1.3.1 Example of a Nonopen Subset of R^2

Figure 1.3.2 An Open Subset of R

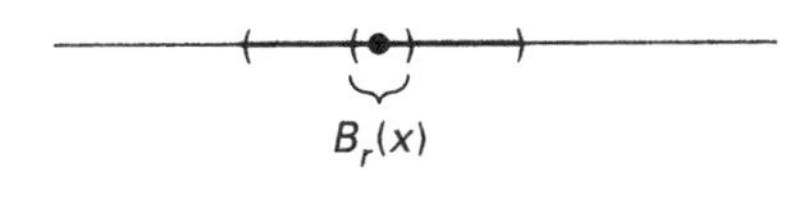

$$[a, \infty) = \{x \in R : x \geq a\},$$
$$(-\infty, b) = \{x \in R : x < b\},$$
$$(-\infty, b] = \{x \in R : x \leq b\}.$$

The adjective "closed" is appropriate for $[a, b]$ because in the following sense it is not possible to get out of the set. If $x_1, x_2, \ldots$ is a sequence of points in $[a, b]$ and $x_n \to x$, then x must also be in $[a, b]$. In fact, this is the general definition of a closed set in a metric space:

If E is a subset of the metric space Ω, E is said to be *closed* if whenever $x_1, x_2, \ldots$ is a sequence of points in E converging to the point $x \in \Omega$, we must have $x \in E$.

Some familiar examples of closed sets in R^2 are the closed balls $C_r(x, y) = \{(x', y') : (x' - x)^2 + (y' - y)^2 \leq r^2\}$ and the closed rectangular boxes $\{(x, y) : a \leq x \leq b,\ c \leq y \leq d\}$. In R, the interval $(0, 1]$ is not closed because the sequence of numbers $1/n$ converges to 0, which is outside the interval. (There are sequences (such as $x_n = 1/2 + 1/n$, $n = 2, 3, \ldots$) in $(0, 1]$ that converge to a limit in $(0, 1]$, but not *every* sequence in $(0, 1]$ converging to a limit has its limit in $(0, 1]$.) Note also that $(0, 1]$ is not open because an open interval centered at 1 cannot be a subset of $(0, 1]$. Intuitively, a closed set is one that contains every one of its boundary points; in this case, the boundary point 0 is outside the set.

The following relation between open and closed sets is basic.

1.3.2 THEOREM. *A set E is open if and only if its complement E^c is closed.*

Proof. Suppose E is open. If $x_1, x_2, \ldots$ is a sequence of points in E^c, and $x_n \to x$, we must show $x \in E^c$. Assume, on the contrary, that $x \in E$. Since E is open, there is an open ball $B_r(x)$ entirely contained in E, and since $x_n \to x$, we have $x_n \in B_r(x)$ for all sufficiently large n (see Fig. 1.3.3(a)). But this contradicts the assumption that all x_n belong to E^c.

Figure 1.3.3 Proof of Theorem 1.3.2

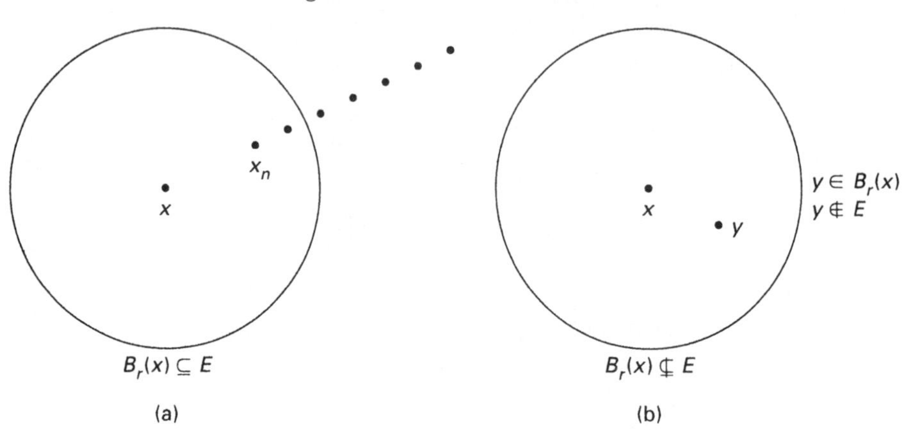

Now assume E^c closed but E not open. To say that E is open means that every point $x \in E$ has an open ball $B_r(x) \subseteq E$, so E not open means that not every point in E has this property. If x is a "bad point" of E, then no open ball $B_r(x)$ is a subset of E; thus, for every $r > 0$, $B_r(x) \not\subseteq E$. (The symbol $\not\subseteq$ means "is not a subset of"; thus, we can find a point $y \in B_r(x)$ with $y \notin E$ (see Fig. 1.3.3(b).) Now if we take $r = 1/n$ and select $x_n \in B_r(x)$ with $x_n \notin E$, we have $d(x_n, x) < 1/n \to 0$, so $x_n \to x$. But $x_n \in E^c$, a closed set, so $x \in E^c$, contradicting the assumption that $x \in E$. ■

Problems for Section 1.3

1. Determine whether the following sets are open in R, closed in R, or neither.

   (a) $\{x \in R : x \leq 0\}$

   (b) $\{x \in R : x < 0 \text{ or } x > 1\}$

   (c) $\{x \in R : x < 0 \text{ or } x \geq 1\}$

   (d) $\{x \in R : x^2 - 3x + 2 < 0\}$

   (e) $[4, \infty)$

2. Repeat Problem 1 for the following subsets of R^2.

   (a) $\{(x, y) : |x| + |y| < 1\}$

   (b) $\{(x, y) : 0 \leq y \leq e^x\}$

   (c) $\{(x, y) : \max(x, y) = 3\}$

   (d) $\{(x, y) : 0 < y \leq e^x\}$

3. You may have the impression that a set cannot be both open and closed, but this is not the case. Show that R is both open and closed in R (similarly, R^2 is both open and closed in R^2).
4. Give an example of another set that is both open and closed in R.
5. Let $\Omega = \{x \in R : x \leq 0 \text{ or } x \geq 1\}$, $E = \{x \in R : x \geq 1\}$. Although E is not an open subset of R, show that E is open (as well as closed) *in* Ω (we will not worry about this situation until later, when we talk about connected sets).
6. Show that any finite set is closed.
7. Show that limits are unique; i.e., if $x_n \to a$ and $x_n \to b$ then $a = b$.
8. Prove the "squeeze theorem"; i.e., if $\{a_n\}$ and $\{b_n\}$ are sequences of real numbers converging to the same limit L, and $a_n \leq x_n \leq b_n$ for all n, then $x_n \to L$ also.

1.4 MINICOURSE IN BASIC LOGIC

Mathematical reasoning sometimes depends on the logical structure of the particular statements and not on their mathematical content. The statements might come from analysis, algebra, geometry, or other areas, but the underlying reasoning process is still the same. In this section we look at this reasoning process.

1.4.1 Truth Tables

One very common type of mathematical statement is the proposition, which has a definite truth value (true (T) or false (F)). New propositions can be created from old ones by using the connectives "and," "or," "not," and "implies." The truth or falsity of the new proposition can be determined by a (probably familiar) device called a *truth table*:

A	B	A and B	A or B	not A
T	T	T	T	F
T	F	F	T	F
F	T	F	T	T
F	F	F	F	T

Thus,

"A and B" is T iff A and B are both T
"A or B" is T iff A or B (*or both*) are T
"not A" is T iff A is F

The "implies" connective is more complicated[2]:

A	B	$A \to B$
T	T	T
T	F	F
F	T	T
F	F	T

Thus, "A implies B" is F only when A is T and B is F

It may seem strange to regard $A \to B$ as true ("by default") when A is F, but here is a supporting argument. Consider the following statement about positive integers:

If n is divisible by 4 then n is divisible by 2. If you agree that this is a true statement, then the truth table for $A \to B$ is forced. Take $n = 5$ to get the FF case, and $n = 6$ to get the FT case.

Another view: the only way to disprove $A \to B$ is to produce a situation in which A is true but B is false.

We now have a formal justification that $\emptyset \subseteq A$ for any set A; i.e., if $x \in \emptyset$, then $x \in A$. The condition "$x \in \emptyset$" is always false, so "if $x \in \emptyset$ then $x \in A$" is *vacuously true* or *true by default*.

1.4.2 Types of Proof

Two very common ideas that occur in the construction of proofs are *proof by contradiction* and *contrapositive*.

[2]If there is any chance of confusing the assertion "A implies B" with a statement about limits, we will use $A \Rightarrow B$ instead of $A \to$B.

Proof by contradiction. If we wish to prove A, we assume (on the contrary) that "not A" holds; i.e., A is false. If we reach a contradiction, then the assumption "not A" is untenable, so A must hold.

Example. See both parts of the proof of Theorem 1.3.2.

Contrapositive. To prove $A \rightarrow B$, we may instead prove the equivalent proposition not $B \rightarrow$ not A, the so-called "contrapositive" of $A \rightarrow B$. The two propositions are equivalent because they have the same truth table:

A	B	not B	not A	not $B \rightarrow$ not A
T	T	F	F	T
T	F	T	F	F
F	T	F	T	T
F	F	T	T	T

Example. Suppose A_1, A_2, and S are sets and $A_1 \subseteq A_2$. If S is not a subset of A_2, then S is not a subset of A_1. The contrapositive is easier to handle: it says that if S is a subset of A_1, then S is a subset of A_2.

1.4.3 Quantifiers

There are mathematical statements or formulas that do not have a definite truth value. For example, suppose we are working in R, and consider the statement

$$x + 3 > 10.$$

This is true for certain values of x, namely, those numbers > 7, but false for all other x.

Similarly, in R^2, the statement $x^2 + y^2 < 1$ holds for certain values of (x, y) and not for others. Thus the "property" $x^2 + y^2 < 1$ defines a subset of R^2, sometimes called a *predicate*, where the property holds.

One way to convert a statement containing variables (such as x and y) into a sentence with a definite truth value is to use *quantifiers*, which are of two types:

> the *universal quantifier* $\forall$ (read "for all"),
> and the *existential quantifier* $\exists$ (read "there exists").

For example, if x stands for a real number, then $\exists x\,(x+3>10)$ is true because there is an x such that $x+3>10$ (any $x>7$ will do). But $\forall x\,(x+3>10)$ is false, since it is definitely not the case that for every $x, x+3>10$.

Both types of quantifiers can appear in the same sentence. For example, $\forall x\,\exists y\,(x+y=10)$ is true since for every x there is a y such that $x+y=10$, namely, $y=10-x$. But $\exists x\,\forall y\,(x+y=10)$ is false; it says that there is an x such that for every $y, x+y=10$. This certainly cannot hold for a fixed x and all possible y.

1.4.4 Mathematical Induction

Let's look at another proof of Theorem 1.2.1: There are 2^n subsets of $\{1,2,\ldots,n\}$. When $n=1$, we are saying that there are two subsets of $\{1\}$, and this is correct; we have $\{1\}$ itself and the empty set. Now if we *assume* that $\{1,2,\ldots,n\}$ has 2^n subsets, we can prove that $\{1,2,\ldots,n+1\}$ has 2^{n+1} subsets, as follows.

Let A be any subset of $\{1,2,\ldots,n+1\}$. There are two mutually exclusive possibilities:

Case 1. $n+1$ does not belong to A. Then A is a subset of $\{1,2,\ldots,n\}$, so there are 2^n such A's.

Case 2. $n+1$ belongs to A. Then A consists of $n+1$ plus a subset of $\{1,2,\ldots,n\}$, so again there are 2^n such A's.

The total number of subsets of $\{1,2,\ldots,n+1\}$ is therefore $2^n+2^n=2^{n+1}$, as asserted.

How do we know that our result is valid for all n? We have verified the case $n=1$, and we know that if it holds for $n=1$, then it holds for

$n = 2$; thus, we have the case $n = 2$. If the theorem is true for $n = 2$, then it is true for $n = 3$, so we have the case $n = 3$. This "domino effect" will eventually reach any positive integer n.

Our proof uses the technique of *mathematical induction,* which may be expressed formally as follows.

For each positive integer n, let $P(n)$ be a statement. A proof by induction that $P(n)$ holds for all n consists of the following steps:

Basis. $P(1)$.

Inductive Step. If $P(n)$, then $P(n+1)$.

Conclude that $P(n)$ *is true for all* n.

In a proof by *strong induction*, the inductive step is replaced by the statement: If $P(1), \ldots, P(n)$ all hold, then so does $P(n+1)$ (the conclusion is the same).

Example (Well–Ordering Principle). If A is a nonempty set of positive integers, then A has a smallest element.

Proof. Suppose that A has no smallest element; show $A = \emptyset$. Here, $P(n)$ is "$n \notin A$". Now, $1 \notin A$, for if $1 \in A$ then 1 is the smallest element of A. If $1 \notin A, 2 \notin A, \ldots, n \notin A$, then $n+1 \notin A$, for if $n+1 \in A$, then $n+1$ is the smallest element of A.

We will encounter only a few proofs by induction (see Section 2.4, Problem 6, and Review Problem 4 in Chapter 2), but there will be many examples of *inductive procedures*. Most commonly, we will generate a sequence $x_1, x_2, \ldots$ where x_n depends on $x_1, \ldots, x_{n-1}$.

1.4.5 Negations

The second part of the proof of Theorem 1.3.2 involves a basic reasoning process in abstract mathematics: going from a statement to its negative. Perhaps the intuition may be aided by the following mechanical procedure, using quantifiers. The statement that E is open means for all $x \in E$ there exists a real number $r > 0$ such that $B_r(x) \subseteq E$; that is,

$$(\forall x \in E)(\exists r > 0)(B_r(x) \subseteq E).$$

Now if we have a mathematical statement of the form $(\forall x)P(x)$ (for every x, $P(x)$ is true), the negation is $(\exists x)$ (not $P(x)$); in other words, for some x, $P(x)$ is false. Similarly, the negation of $(\exists x)P(x)$ is $(\forall x)$ (not $P(x)$). Thus, to obtain the negation, we simply reverse all the quantifiers and change the statement on the extreme right of the expression to its negative. Therefore, "E is not open" means

$$(\exists x \in E)(\forall r > 0)(B_r(x) \not\subseteq E);$$

there is an $x \in E$ such that for every $r > 0$, the open ball $B_r(x)$ is not a subset of E, just as we found in the proof of Theorem 1.3.2.

Note. In a phrase such as $\forall x \in E$, the fragment "$\in E$" refers to the range of the variable x and is *not changed* when taking the negative.

Problems for Section 1.4

1. The *converse* of $A \to B$ is $B \to A$. Give an example in which $A \to B$ is true but $B \to A$ is false.
2. Verify the De Morgan laws for propositions; i.e.,

   not (A and B) is equivalent to (not A) or (not B);

   not (A or B) is equivalent to (not A) and (not B).

3. When we say that P and Q are equivalent, we mean that P and Q have the same truth value regardless of the truth or falsity of the component propositions that make up P and Q. (See the discussion of contrapositives in the text.) Show that this amounts to the assertion that the proposition "P if and only if Q," i.e., $(P \to Q)$ and $(Q \to P)$, is true.

   notation $P \longleftrightarrow Q$

4. Suppose x and y range over the positive real numbers (i.e., $x, y > 0$). Determine whether the following statements are true or false:

   (a) $\forall x \, \exists y \, (x < y)$

   (b) $\exists x \, \forall y \, (x \leq y)$

5. Write the statement $\lim_{n \to \infty} d(x_n, x) = 0$ using universal and existential quantifiers.

6. Use the technique given in the text to take the negation of the statement of Problem 5. Then express your result in ordinary language.
7. Suppose we have proved "if P then not P." Show that we do not necessarily have a contradiction, but we can conclude "not P."

1.5 LIMIT POINTS AND CLOSURE

The following problem frequently arises in analysis: we have a point x and a set E, and for each open ball $B_r(x)$ we wish to select, if possible, a point $y \in B_r(x) \cap E$. If we can do this for every $r > 0$, and furthermore, the point y can be chosen to be unequal to x, then x is called a *limit point* or *cluster point* of E. Formally, x is a limit point of E if every open ball $B_r(x)$ contains a point $y \in E$ with $y \neq x$. Thus there are points in E arbitrarily close to x (not counting x itself); i.e., every *deleted* open ball about x ($B_r(x)$ with x removed) contains a point of E. For example, let $E = (0, 1] \cup \{2\} = \{x \in R : 0 < x \leq 1 \text{ or } x = 2\}$ (see Fig. 1.5.1). If $0 \leq x \leq 1$, every open interval $(x - r, x + r)$ has a point of E other than x, so that every point in $[0, 1]$ is a limit point of E. (Note that a limit point of E need not belong to E, for example, the point 0.) An *isolated point* of E is a point *belonging to* E that is not a limit point. In Fig. 1.5.1, E has one isolated point, namely 2.

If x is a limit point of E, we can find points $x_n \in B_{1/n}(x)$ with $x_n \neq x$. Thus, x is the limit of a sequence of points $x_n \in E$ with x_n never equal to x. We can now describe precisely those points that can be expressed as limits of sequences from E.

1.5.1 THEOREM. *Suppose E is a subset of the metric space Ω. Let E' be the set of limit points of E, and define $\bar{E} = E \cup E'$; $\bar{E}$ is called the closure of E. Then*

(a) The point x belongs to $\bar{E}$ if and only if there is a sequence of points $x_n \in E$ with $x_n \to x$.
(b) $\bar{E}$ is a closed set.

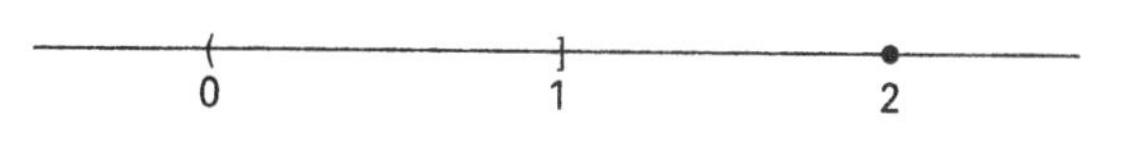

Figure 1.5.1 Limit Points and Isolated Points

(c) If C is a closed set and $E \subseteq C$, then $\bar{E} \subseteq C$; thus, $\bar{E}$ is the smallest closed set containing E.
(d) E is closed if and only if $E = \bar{E}$.

Proof

(a) Let $x \in \bar{E}$. If $x \in E$, take $x_n = x$ for all n; if $x \in E'$, then there are points $x_n \in E$ with $x_n \to x$ (and x_n never $= x$).
Conversely, if $x_n \in E$ and $x_n \to x$, we have two possibilities (a *proof by cases*). If $x \in E$, then $x \in \bar{E}$ by definition of $\bar{E}$; if $x \notin E$, then any open ball $B_r(x)$ will contain the x_n for all sufficiently large n (because $x_n \to x$). Since $x_n \in E$, we have $x_n \neq x$, so $x \in E' \subseteq \bar{E}$.
(b) Let $x_n \in \bar{E}$, $x_n \to x$; we must show $x \in \bar{E}$. Assume the contrary, that is, $x \notin E$ and $x \notin E'$. Then for some $r > 0$, $B_r(x)$ contains no points of E. But $x_n \in B_r(x)$ for all sufficiently large n; consider any such x_n (Fig. 1.5.2). Since $x_n \in B_r(x)$, we have $x_n \notin E$, and since $x_n \in \bar{E}$ we must have $x_n \in E'$. Choose $s > 0$ so small that $B_s(x_n) \subseteq B_r(x)$, and select $z \in B_s(x_n)$ such that $z \in E$. This contradicts the fact that $B_r(x)$ has no points of E.
(c) If $x \in \bar{E}$, then by (a) we have a sequence of points $x_n \in E$ with $x_n \to x$. By hypothesis, $E \subseteq C$; therefore $x_n \in C$. Also by hypothesis, C is closed; therefore $x \in C$.
(d) If $E = \bar{E}$, then E is closed by (b). Conversely, if E is closed, then by (c) (with $C = E$) we have $\bar{E} \subseteq E$. But E is always a subset of $\bar{E}$, so $E = \bar{E}$. ■

Intuitively, $\bar{E}$ is E together with all of its boundary points.

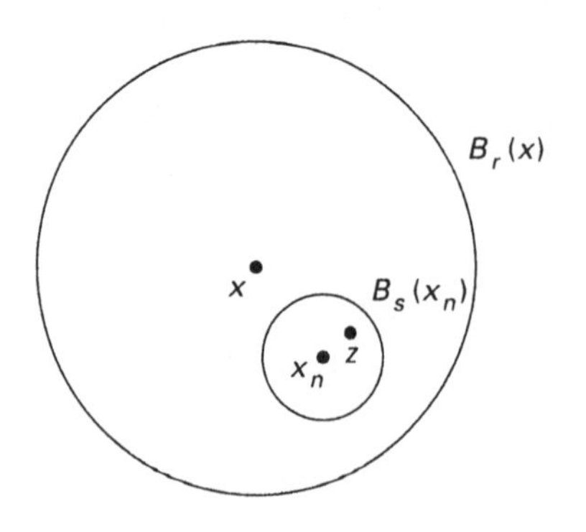

Figure 1.5.2 Proof of Theorem 1.5.1

Problems for Section 1.5

1. In the text, we showed that if x is a limit point of E then there is a sequence of points $x_n \in E$ with x_n never equal to x. Show that, conversely, if $x_n \in E$, x_n never equal to x, and $x_n \to x$, then x is a limit point of E.
2. Let $E = \{(x,y) : 0 \le x < 1, 0 < y \le 1\}$. Find E' and $\bar{E}$.
3. In R, let $E = \{1, \frac{1}{2}, \frac{1}{3}, \frac{1}{4}, \dots\}$. Find E' and $\bar{E}$.
4. If E' is the set of limit points of E, show that E' is closed.
5. Show that x is a limit point of E if and only if x is a limit point of $\bar{E}$.
6. Let E' be the set of limit points of E. If x is a limit point of E', show that x is a limit point of E.
7. Let $E = (0,1) \cup \{2\frac{1}{2}, 2\frac{1}{3}, 2\frac{1}{4}, \dots, 2 + \frac{1}{n}, \dots\}$.
   (a) Find E'.
   (b) Exhibit a limit point of E that is not a limit point of E'.
8. We may give a formal definition of boundary, as follows: x is a *boundary point* of the set E if every open ball $B_r(x)$ contains both a point of E and a point of E^c; x is an *interior point* of E if there is an $r > 0$ such that $B_r(x) \subseteq E$. Show that x is a boundary point of E iff $x \in \bar{E}$ and x is not an interior point of E.

REVIEW PROBLEMS FOR CHAPTER 1

1. Call a set of positive integers *co-finite* if its complement is finite. For example, $\{1,3,4,8,9,10,11,\dots\}$ is co-finite because its complement $\{2,5,6,7\}$ is finite. Are there countably or uncountably many co-finite subsets of the positive integers? Justify your answer.
2. Consider the following statement: If x is a limit point of $\bar{E}$, then x is a limit point of E. If the statement is true, explain why. If it is false, give an explicit counterexample.
3. We defined an *open ball* as a set of the form

$$B_r(x) = \{y \in \Omega : d(x,y) < r\}, \qquad r > 0.$$

Prove formally (without appealing to a picture) that $B_r(x)$ is actually an open set.

4. Consider the following mathematical statement involving quantifiers:

$$(\forall \epsilon > 0)(\exists N)(\forall n, m \geq N)(d(x_n, x_m) < \epsilon).$$

(This says that $\{x_n\}$ is a *Cauchy sequence*; we study such sequences in Section 2.4).

   (a) Write the negation of the statement (again using quantifiers).

   (b) If the negation holds, show that for some $\epsilon > 0$, there are distinct integers $n_1, m_1, n_2, m_2, \ldots$ such that $d(x_{n_k}, x_{m_k}) \geq \epsilon$ for all k.

5. Give an example of a nonempty set of real numbers with no limit point.

2

SOME BASIC TOPOLOGICAL PROPERTIES OF R^p

2.1 UNIONS AND INTERSECTIONS OF OPEN AND CLOSED SETS

Properties that can be expressed in terms of open and closed sets are called *topological*. In this chapter, we develop results of this type. First, we need to know if we can take unions and intersections of open sets and still remain within the class of open sets (a similar question arises for closed sets). Suppose, for example, that A_1, A_2, and A_3 are open (all sets are subsets of a fixed metric space Ω); is $A_1 \cup A_2 \cup A_3$ open? If x belongs to $\bigcup_i A_i$, then x is in A_i for some i, and we can find an open ball $B_r(x)$ contained in A_i (see Fig. 2.1.1). But then $B_r(x)$ is a subset of $\bigcup_i A_i$, proving that $A_1 \cup A_2 \cup A_3$ is open. In fact, this argument works for an arbitrary collection of open sets, and we have our first result.

2.1.1 THEOREM. *An arbitrary union of open sets is open. Thus, if A_i is open for all i (where the A_i's can form a finite collection, a countably infinite collection, or even an uncountably infinite collection), then $\bigcup_i A_i$ is open.*

Now let's consider the same question for intersections of open sets. If x belongs to $A_1 \cap A_2 \cap A_3$, where A_1, A_2, and A_3 are open sets, we

Figure 2.1.1 An Arbitrary Union of Open Sets Is Open

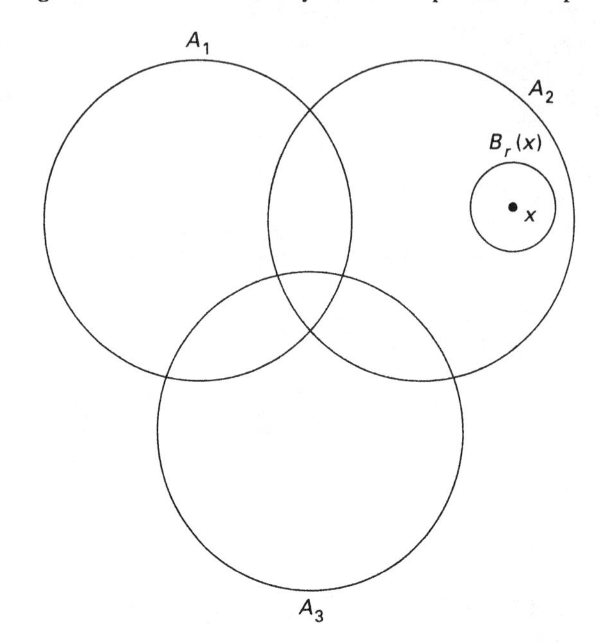

can find open balls $B_{r_1}(x)$, $B_{r_2}(x)$, and $B_{r_3}(x)$ with $B_{r_j}(x) \subseteq A_j$ for each j. If we simply take the smallest radius, that is, $r = \min_j r_j$, then $B_r(x) \subseteq A_1 \cap A_2 \cap A_3$. This gives us the second result (see Fig. 2.1.2).

2.1.2 THEOREM. *A finite intersection of open sets is open. That is, if n is a positive integer and $A_1, \ldots, A_n$ are open sets, then $\bigcap_{i=1}^{n} A_i$ is open.*

Why doesn't the above argument work when there are infinitely many sets? Simply because the "smallest radius" r might be 0; for example, we might have $r_j = 1/j, j = 1, 2 \ldots.$ Now if a particular argument fails, we cannot conclude that a result is invalid. However, in this case we can give an *explicit example* to show that Theorem 2.1.2 is false for infinite intersections. Let A_n be the open interval $(-1/n, 1/n)$ in R. Then $\bigcap_{n=1}^{\infty} A_n$ is the set of real numbers x such that $-1/n < x < 1/n$ for every positive integer n; thus $\bigcap_{n=1}^{\infty} A_n = \{0\}$, the set consisting of $\{0\}$ alone, which is not open.

Caution. The notation $\bigcap_{n=1}^{\infty} A_n$, although standard, may be confusing. It means the set of points x such that x belongs to A_n for

Figure 2.1.2 A Finite Intersection of Open Sets Is Open

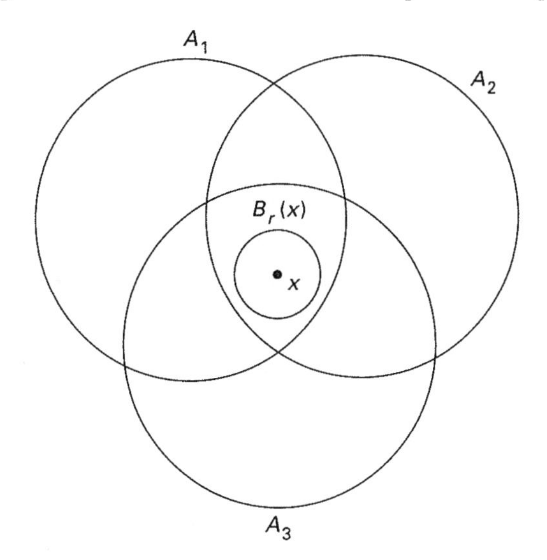

every positive integer n. There is no set A_∞; in other words, *infinity is not a positive integer.*

To obtain appropriate results for sets that are closed rather than open, we can use the De Morgan laws and Theorem 1.3.2.

2.1.3 THEOREM

(a) An arbitrary intersection of closed sets is closed. In other words, if B_i is closed for all i, then $\bigcap_i B_i$ is closed.
(b) A finite union of closed sets is closed. That is, if n is a positive integer and $B_1, \ldots, B_n$ are closed sets, then $\bigcup_i B_i$ is closed.

Proof

(a) By the De Morgan laws, $(\bigcap_i B_i)^c = \bigcup_i B_i^c$, which is open by Theorems 1.3.2 and 2.1.1. Thus, by Theorem 1.3.2, $\bigcap_i B_i$ is closed.
(b) Again by the De Morgan laws, $(\bigcup_{i=1}^n B_i)^c = \bigcap_{i=1}^n B_i^c$, which is open by Theorems 1.3.2 and 2.1.2. By Theorem 1.3.2, $\bigcup_{i=1}^n B_i$ is closed. ■

Problems for Section 2.1

1. Give an example of an infinite union of closed sets that is not closed.
2. Show that an open subset of a metric space can be expressed as a union of open balls.

   The following problems will help to strengthen the result of Problem 2 later, in the special case when the metric space is R.
3. Let V be an open subset of R. If $x \in V$, let V_x be the union of all open intervals I such that $x \in I$ and $I \subseteq V$. For example, if $V = (0,1) \cup (2,3) \cup (4,5)$ and $x = 2.7$, then $V_x = (2,3)$, the largest open interval containing 2.7 and contained in V.

   Later, we show that V_x is always an open interval, but let's assume this result for now. If $x \neq y$, show that V_x and V_y are either disjoint or identical.
4. Continuing Problem 3, show that there are only countably many distinct V_x's. (The counting technique used here is very basic. Form a set S by choosing an element from each distinct V_x. Since the distinct V_x are disjoint, all the elements of S are distinct. If S turns out to be countable, there can only be countably many distinct V_x.)

2.2 COMPACTNESS

When the real numbers are defined by a set of axioms, one property of a topological nature is built right into the axioms, and the other topological properties are then derived. There is a wide choice of the particular property to assume. Our choice will be somewhat unorthodox and is motivated by our desire to keep proofs as simple as possible.[1] Eventually we consider the standard approach and examine its relationship to ours. Of course, if the reals are obtained by a constructive procedure, then we don't have to assume anything, but at some point we have to choose how to construct the real numbers, and the choice is

[1] Despite our best efforts, the proofs in this section may still be difficult, especially the Heine–Borel theorem. If you have trouble with the details, don't worry, it's normal. Concentrate on the explicit examples and on understanding the definitions, and you should be able to cope with the applications later.

made so that all the properties we regard as desirable can be obtained. First, some terminology.

2.2.1 Definition

The set B is said to be *bounded* if it can be enclosed by a ball, that is, if $B \subseteq B_r(x)$ for some $x \in \Omega$ and $r > 0$.

We now state our assumption (due to Cantor).

2.2.2 Nested Set Property

If $B_1, B_2, \ldots$ is a sequence of closed, bounded, nonempty subsets of R^p, and the sequence is *nested*, that is, $B_{n+1} \subseteq B_n$ for all $n = 1, 2, \ldots,$ then $\bigcap_{n=1}^{\infty} B_n$ is not empty.

Property 2.2.2 probably looks obscure, but if we examine closed intervals in R the result should appear more reasonable. Suppose $B_j = [a_j, b_j], j = 1, 2, \ldots.$ The nesting requirement means that the B_j are shrinking as j increases, in other words, the a_j will increase (not necessarily strictly) and the b_j will decrease. It is reasonable to expect that a_j will approach a limit a, and b_j will converge to b. Since $a_j \leq b_j$ for all j, we have $a \leq b$; consequently, $\bigcap_{n=1}^{\infty} B_n = [a, b] \neq \emptyset$.

Note that the Nested Set Property does not apply to *open* sets. For example, in R the sets $A_n = (0, 1/n)$ form a nested sequence, but $\bigcap_{n=1}^{\infty} A_n = \emptyset$.

Whenever we use the words "increase" and "decrease," as above, the strict connotation is *not* implied.

Now consider the following problem. If x is a point of the interval $[0, 1]$, let G be the open interval $(x - 1/4, x + 1/4)$. The sets G_x *cover* [0,1]; that is, $[0, 1] \subseteq \bigcup_{0 \leq x \leq 1} G_x$. Do we actually need all the G_x in order to form a covering? In fact $G_{.2}$, $G_{.5}$, and $G_{.8}$ are sufficient, since any x in $[0, 1]$ will belong to at least one of these three sets. In this way, we have reduced the given open covering to a finite subcovering. If a set has the property that such a reduction can always be carried out, it is called compact.

2.2.3 Definition

Let K be a subset of the metric space Ω; K is *compact* if every open covering of K has a finite subcovering; that is, if $K \subseteq \bigcup_i G_i$, where the G_i are open sets, then $K \subseteq G_{i_1} \cup G_{i_2} \cup \cdots \cup G_{i_n}$ for some $i_1, \ldots, i_n$.

It will be easy to recognize compact sets in R^p; they are precisely the sets that are closed and bounded. First we show that a compact set in *any* metric space is always closed and bounded.

2.2.4 THEOREM. *If K is a compact subset of the metric space Ω, then K is closed and bounded.*

Proof. To show K closed, we prove that K^c is open. Assume $x \notin K$, and let $G_m = \{y : d(x,y) > 1/m\}$, $m = 1, 2, \ldots$. If $y \in K$, then $x \neq y$; hence, $d(x,y) > 1/m$ for some m; therefore $y \in G_m$ (see Fig. 2.2.1). Thus, $K \subseteq \bigcup_{m=1}^{\infty} G_m$, and by compactness we have a finite subcovering. Now observe that the G_m form an *increasing sequence* of sets ($G_1 \subseteq G_2 \subseteq G_3 \subseteq \cdots$); therefore, a finite union of some of the G_m, for example $G_3 \cup G_7 \cup G_{11}$, is equal to the set (G_{11}) with the highest index. Thus, $K \subseteq G_s$ for some s, and it follows that $B_{1/s}(x) \subseteq K^c$. (If $d(x,y) < 1/s$, then $y \notin G_s$ so $y \notin K$.) Therefore, K^c is open.

To show K is bounded, fix $r > 0$. If $x \in K$, then $x \in B_r(x)$, so K is covered by the $B_r(x)$; by compactness, we have a finite subcovering, say $K \subseteq B_r(x_1) \cup \cdots \cup B_r(x_n)$. But a finite union of open balls can be placed inside a single open ball, so K is bounded. ∎

We now show that closed, bounded subsets of R^p are compact.

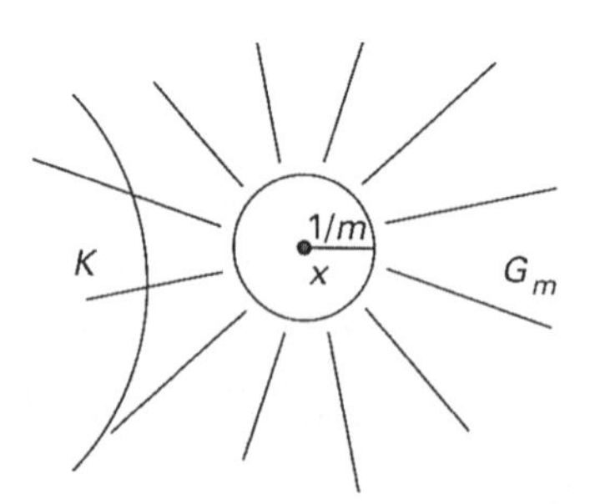

Figure 2.2.1 Proof That a Compact Set Is Closed

2.2.5 Heine-Borel Theorem

If K is a subset of R^P, then K is compact if and only if K is closed and bounded.

Proof. In view of Theorem 2.2.4, we need only show that if K is closed and bounded, it must be compact. Assume $K \subseteq \bigcup_i G_i$, with all G_i open. If $x \in K$, then $x \in G_i$ for some i, and since G_i is open there is an open ball $B_r(x) \subseteq G_i$. Now there is a rational number y in R^P (this means $y = (y_1, \ldots, y_p)$, where each y_i is rational) and a rational number $s > 0$ such that $x \in B_s(y)$ and $B_s(y) \subseteq B_r(x)$: see Fig. 2.2.2. Let's write down the relevant information in a table with three (very long) columns:

$$x \qquad G_i \qquad B_s(y).$$

The first column contains the points x in K; the second and third columns contain specific choices of G_i and $B_s(y)$. It might happen that a particular $B_s(y)$ appears more than once on the list, with a different G_i; if so, toss out the new G_i and replace it by the original. Since there are only countably many sets $B_s(y)$, there will be only countably many G_i, in other words, K will be covered by a countable union of the G_i. The fact that K is closed and bounded has not yet been used; we have shown that *if K is an arbitrary subset of R^P an open covering of K has a countable subcovering.*

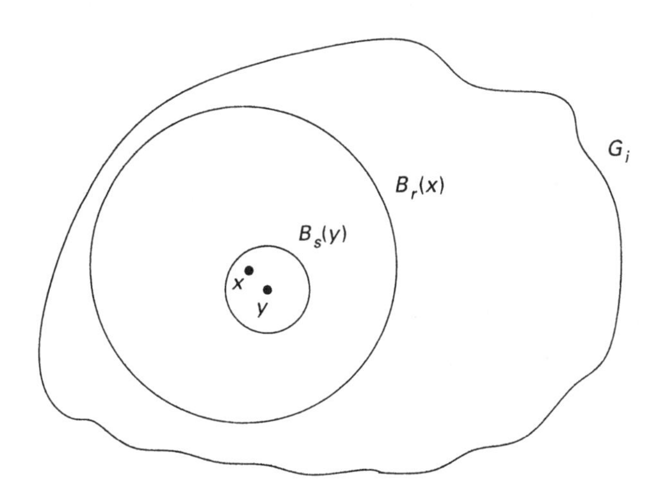

Figure 2.2.2 Proof of the Heine-Borel Theorem

It remains to reduce the countable covering to a finite subcovering. Assume $K \subseteq \bigcup_{j=1}^{\infty} G_j$; if $H_n = G_1 \cup \cdots \cup G_n$, then the H_n are open and $H_1 \subseteq H_2 \subseteq H_3 \subseteq \cdots$ and $K \subseteq \bigcup_{n=1}^{\infty} H_n$. Let $B_n = H_n^c \cap K$; since H_n^c and K are closed and K is bounded, the B_n are closed and bounded. Furthermore, $B_{n+1} \subseteq B_n$ (because $H_n \subseteq H_{n+1}$, and therefore $H_{n+1}^c \subseteq H_n^c$). If all B_n are nonempty, the Nested Set Property implies that there is a point $x \in \bigcap_{n=1}^{\infty} B_n$. By definition of B_n we have $x \in K$, and for all n, $x \in H_n^c$; that is, $x \notin H_n$. This contradicts the fact that $K \subseteq \bigcup_{n=1}^{\infty} H_n$.

It follows that for some m, $B_m = \emptyset$. Now if $y \in K$, then y must be in H_m; otherwise $y \in K \cap H_m^c = B_m$. Thus, $K \subseteq H_m = \bigcup_{j=1}^{m} G_j$, a finite subcovering. ■

The following examples may aid the intuition. If $A = (0, 1)$, then A is bounded but not closed, and, therefore, is not compact. In fact, if $G_n = (0, 1 - 1/n)$, $n = 1, 2, \ldots$, then $A = \bigcup_{n=1}^{\infty} G_n$, but there is no finite subcovering. If $\epsilon > 0$ and $H_1 = (-\epsilon, \epsilon)$, $H_2 = (1 - \epsilon, 1 + \epsilon)$, then the G_n together with H_1 and H_2 cover the compact set $[0, 1]$, but there is a finite subcovering. (We can use H_1, H_2 and one of the G_n if $1/n < \epsilon$.)

The sets $(-r, r)$, r real and greater than zero, cover R, but there is no finite subcovering. However, there is a *countable* subcovering: $R = \bigcup_{n=1}^{\infty}(-n, n)$.

If Z is the set of integers, then Z is closed (Z^c is open) but unbounded, and thus not compact. Explicitly, if $G_n = (n - a, n + a)$, where $0 < a < 1/2$, $n = 1, 2, \ldots$, the G_n form an open covering of Z with no finite subcovering.

Problems for Section 2.2

1. Which of the following subsets of R are compact?
   (a) $(-\infty, 3)$
   (b) $(-7, 3]$
   (c) $[-7, 3]$
   (d) $[-7, 3)$
   (e) $[3, \infty)$

2. The set $C = \{x \in R : x \geq 1\}$ is closed but unbounded and, hence, not compact. Given an explicit example of an open covering of C with no finite subcovering.
3. Let $D = \{(x,y) : x^2 + y^2 < 1\}$, a bounded but not closed, and hence not compact, subset of R^2.
   (a) Give an example of an open covering of D with no finite subcovering.
   (b) Give an example of an open covering of D that *does* have a finite subcovering.
4. Why doesn't the example of Problem 3(b) contradict the noncompactness of D?
5. Is the empty set compact?

2.3 SOME APPLICATIONS OF COMPACTNESS

Compactness is closely related to convergence of subsequences. It may be useful at this point to give the formal definitions of sequences and subsequences. A *sequence* $x_1, x_2, \ldots$ in a metric space Ω is really a *function* from the positive integers to Ω, in other words, for each positive integer n we have a point x_n in Ω. To form a *subsequence* of this sequence, we assign to each positive integer i a point x_{n_i}, subject to the requirement that the indices are strictly increasing; that is, $i < j$ implies $n_i < n_j$. In other words, $x_2, x_6, x_{15}, x_{17}, \ldots$ is allowable but not $x_3, x_9, x_8, x_{14}, \ldots$ or $x_3, x_9, x_9, x_{14}, \ldots$.

Sometimes the notation $\{x_1, x_2, \ldots\}$, or simply $\{x_n\}$, is used both for the sequence and for its set of values, but the mathematical ideas are different. For example, say we have a constant sequence: $x_n = 3$ for all n. The sequence is a function that assigns to each positive integer n the number 3; the set of values of the sequence is the set $\{3\}$ consisting of 3 alone. We'll try not to cause any notational confusion. If we want to use the notation $\{x_n\}$ for a sequence, we'll say "the sequence $\{x_n\}$" to make it clear that we mean the sequence, not its set of values.

A sequence of points in a compact set always has a convergent subsequence.

2.3.1 THEOREM. *Let K be a compact subset of the metric space Ω. If $x_1, x_2, \ldots$ is a sequence of points in K, there is a subsequence $x_{n_1}, x_{n_2}, \ldots$ converging to a limit x in K.*

Proof. Fix a point $a \in K$ and consider the following inductive process. Find (if possible) a positive integer n_1 such that $x_{n_1} \in B_1(a)$. If this can be done, then find (if possible) a positive integer $n_2 > n_1$ such that $x_{n_2} \in B_{1/2}(a)$. In general, we try to find positive integers $n_1 < n_2 < \cdots$ such that $x_{n_j} \in B_{1/j}(a), j = 1, 2, \ldots$ (see Fig. 2.3.1). If the process does not terminate, then $x_{n_j} \to a$ and we are finished. If the process terminates at n_i then $B_{1/(i+1)}(a)$ contains no point x_n, $n > n_i$. Thus if $r_a = 1/(i+1)$, then $x_n \in B_{r_a}(a)$ for only finitely many n. Carry out the procedure for each $a \in K$. If for some a the process does not terminate, we are finished, so assume that for each $a \in K$ we get an open ball $G_a = B_{r_a}(a)$ such that $x_n \in B_{r_a}(a)$ for only finitely many n. Since $a \in G_a$, we have $K \subseteq \bigcup_{a \in K} G_a$, and, by compactness, there is a finite subcovering, say $K \subseteq \bigcup_{i=1}^m G_{a_i}$. It follows that $x_n \in K$ for only finitely many n, contradicting the assumption that the entire sequence lies in K. ■

If we specialize Theorem 2.3.1 to R^p, we obtain the following basic result.

2.3.2 Bolzano–Weierstrass Theorem

Let $x_1, x_2, \ldots$ be a bounded sequence in R^p (in other words, the set of values $\{x_1, x_2, \ldots\}$ is a bounded subset of R^p). Then there is a subsequence $x_{n_1}, x_{n_2}, \ldots$ converging to a point x in R^p.

Proof. We may assume that all the x_n belong to a fixed closed

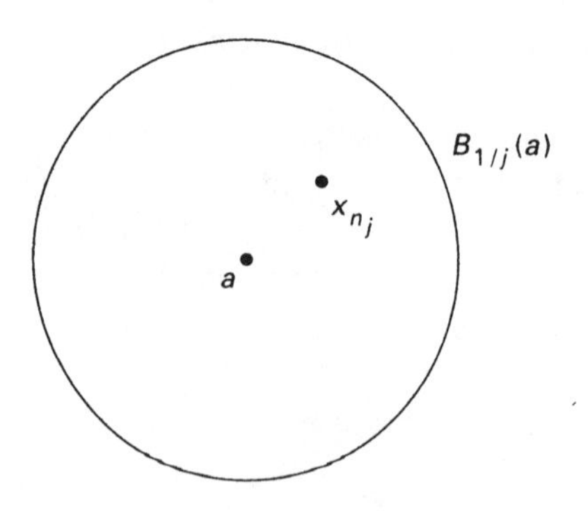

Figure 2.3.1 Proof of Theorem 2.3.1

ball $C_r(a) = \{x : d(x,a) \le r\}$. By the Heine–Borel Theorem 2.2.5, $C_r(a)$ is compact, and the result follows from Theorem 2.3.1. ■

2.3.3 COROLLARY. *If A is a bounded infinite subset of R^p, then A has a limit point.*

Proof. Let $x_1, x_2, \ldots$ be a sequence of *distinct* points in A; such a sequence exists because A is infinite. By Theorem 2.3.2 we have a convergent subsequence $x_{n_i} \to x \in R^p$. Now consider the open ball $B_r(x)$; for large enough i, $x_{n_i} \in B_r(x)$. If $x_{n_i} = x$, then (because the points of the sequence are distinct) $x_{n_{i+1}} \ne x$. Thus, we can find a point $y \in A \cap B_r(x)$ with $y \ne x$. Therefore, x is a limit point of A. ■

If $x_1, x_2, \ldots$ is a convergent sequence in R^p (say $x_n \to x$), the sequence must be bounded, for if $r > 0$ then x_n will lie in the open ball $B_r(x)$ for all sufficiently large n, say for $n \ge N$. This leaves only finitely many points $x_1, \ldots, x_{N-1}$, so if we make s sufficiently large we can enclose the entire sequence in the ball $B_s(x)$. Thus, an unbounded sequence cannot converge. It is sometimes convenient to relax this requirement, especially when working in the set of real numbers R. We can do this by adding $+\infty$ and $-\infty$ to R to obtain the set of *extended real numbers* $\bar{R}$. The rules of order and arithmetic in $\bar{R}$ are natural:

$$-\infty \le x \le \infty \qquad \text{for all} \quad x \in \bar{R},$$

$$a + \infty = \infty + a = \infty, \quad a - \infty = -\infty + a = -\infty, \qquad a \in R,$$

$$\infty + \infty = \infty, \quad -\infty - \infty = -\infty \qquad (\infty - \infty \quad \text{is not defined}),$$

$$b \cdot \infty = \infty \cdot b = \begin{cases} \infty & \text{if } b \in \bar{R}, \quad b > 0, \\ -\infty & \text{if } b \in \bar{R}, \quad b < 0, \end{cases}$$

$$\frac{a}{\infty} = \frac{a}{-\infty} = 0, \quad a \in R \quad \left(\frac{\infty}{\infty} \quad \text{is not defined}\right).$$

We do not define $0 \cdot \infty$, although in some areas of mathematics (notably measure theory), it is required that $0 \cdot \infty = 0$.

When working in $\bar{R}$, we encounter the problem that any point in R is infinitely far from $+\infty$ and $-\infty$. It is possible to modify the Euclidean metric so that we get a legitimate metric on $\bar{R}$ without changing the open sets of R, but this involves more topological equipment than we have available at present. Instead, let's define convergence to $\pm\infty$ directly:

> $x_n \to \infty$ means for each positive real number M, $x_n > M$ for all sufficiently large n; that is, there is a positive integer N such that $x_n > M$ for all $n \geq N$.
>
> $x_n \to -\infty$ means for each negative real number M, $x_n < M$ for all sufficiently large n.

We can now establish an important property of unbounded sequences.

2.3.4 THEOREM. *If $x_1, x_2, \ldots$ is an unbounded sequence in R, there is either a subsequence converging to $+\infty$ or a subsequence converging to $-\infty$. Thus, by Theorem 2.3.2, every sequence in R has a convergent subsequence if $+\infty$ and $-\infty$ are allowed as limits.*

Proof. Since the sequence is unbounded, not all x_n can lie in the interval $(-1, 1)$; hence, $|x_{n_1}| \geq 1$ for some n_1. The sequence x_{n_1+1}, $x_{n_1+2}, \ldots$ is also unbounded, so not all of its points can lie in $(-2, 2)$; consequently, $|x_{n_2}| \geq 2$ for some $n_2 > n_1$. Inductively, we find a sequence $x_{n_1}, x_{n_2}, \ldots$ such that $|x_{n_k}| \geq k$ for all k. Now either $x_{n_i} > 0$ for infinitely many i, or $x_{n_i} < 0$ for infinitely many i. In the first case we can toss out the negative terms to obtain a subsequence converging to $+\infty$; in the second case we remove the positive terms to obtain a subsequence converging to $-\infty$. ■

We conclude this section with a convergence result that will be used many times.

2.3.5 THEOREM. *If $\{x_n\}$ is a sequence of real numbers that converges to the real number L, and $x_n \leq c$ for all n, then $L \leq c$. (Similarly, if $x_n \to L$ and $x_n \geq c$ for all n, then $L \geq c$.)*

Proof. If $L > c$, then, since $x_n \to L$, x_n will be greater than c for all sufficiently large n, a contradiction. ■

Problems for Section 2.3

1. Find the limits, if they exist, of each of the following sequences $\{x_n\}$ in R.

   (a) $\dfrac{n^2+3n-4}{n^3+n^2+1}$

   (b) n^3e^{-n}

   (c) $\arctan n$

   (d) $(1-\dfrac{3}{n})^n$

   (e) $\sin\dfrac{\pi n}{2}$

2. Suppose $x_1, x_2, \ldots \in R^p$, $y \in R^p$, and $d(x_n, y) \leq k$ for all n. If $x_n \to x$, show that $d(x, y) \leq k$.

3. Given an example of a sequence in the noncompact set $(0, 1)$ with no subsequence converging to a point of $(0, 1)$.

4. Consider the extended reals $\bar{R}$ and recall the definition of open set: G is open if and only if, for each $x \in G$, there is an open ball $B_r(x) \subseteq G$. If we try to extend this idea to $\bar{R}$, difficulties arise when $x = +\infty$ or $-\infty$. However, if $x = +\infty$ we can replace $B_r(x)$ by a set of the form $\{y \in \bar{R} : y > r\}$, and if $x = -\infty$ we can use $\{y \in \bar{R} : y < r\}$. With this adjustment, show that every open covering of $\bar{R}$ has a finite subcovering, so $\bar{R}$ is compact.

5. Continuing Problem 4, note that we may use the same definition of a closed subset of $\bar{R}$: C is closed if and only if for every sequence $x_1, x_2, \ldots$ of points of C converging to a limit $x \in \bar{R}$, we must have $x \in C$. Give an example of a set $A \subseteq R$ such that A is closed in R but not in $\bar{R}$; in other words, if we regard R as the universe, A is a closed set, but if we regard $\bar{R}$ as the universe, A is not closed.

6. Continuing Problem 5, is it possible for a set $A \subseteq R$ to be closed in $\bar{R}$ but not in R? Explain.

2.4 LEAST UPPER BOUNDS AND COMPLETENESS

We now examine the problem of finding the maximum of a set of real numbers, a familiar idea from calculus. For example, if E is the interval [0,1], we observe immediately that 1 is the largest element of E. But what if 1 is removed, so that we have a new set $E_1 = [0, 1)$? In this case, E_1 has no largest element, but 1 is the "closest thing to it." Mathematically, we have an example of a least upper bound, which we now define.

2.4.1 Definitions

Let E be a nonempty subset of R. The real number x is said to be an *upper bound* of E if $x \geq y$ for every $y \in E$: similary, x is a *lower bound* of E if $x \leq y$ for every $y \in E$. We say that x is a *least upper bound* or *supremum* (abbreviated *sup*) of E if x is an upper bound that is less than or equal to all other upper bounds of E; similarly, x is a *greatest lower bound* or *infimum* (abbreviated *inf*) of E if x is a lower bound that is greater than or equal to all other lower bounds of E.

Note that if x and y are least upper bounds of E, we must have $x = y$, for if (say) $x < y$, then y cannot be the smallest upper bound. Thus, we may refer without ambiguity to "the" least upper bound. The following existence theorem is basic.

2.4.2 THEOREM. *Let E be a nonempty subset of R. If E has an upper bound, then E has a least upper bound. Similarly, a nonempty subset of R that has a lower bound has a greatest lower bound.*

Proof. Let x_1 be an upper bound of E, and let $y_1 \in E$, so that $y_1 \leq x_1$. Let m be the midpoint of the interval $[y_1, x_1]$ (see Fig. 2.4.1). If m is an upper bound of E, take $x_2 = m, y_2 = y_1$, so that x_2 is an upper bound of E and $y_2 \in E$. If m is not an upper bound of E, then some element of E, say y_2, is greater than m. (Necessarily $y_2 \leq x_1$ since x_1 is an upper bound of E.) Take $x_2 = x_1$, and again x_2 is an upper bound of E and $y_2 \in E$. Inductively, we obtain closed intervals

$$C_n = [y_n, x_n] \quad \text{with} \quad x_n \text{ an upper bound of } E \text{ and } y_n \in E;$$

furthermore, the x_n decrease and the y_n increase, so the C_n form a nested sequence of closed, bounded, nonempty sets. By the Nested

Figure 2.4.1 Proof of Theorem 2.4.2

Set Property 2.2.2, $\bigcap_{n=1}^{\infty} C_n$ contains at least one point x. Since $y_n \le x \le x_n$ and

$$|x_n - y_n| \le (\frac{1}{2})^{n-1}|x_1 - y_1| \to 0,$$

we have $x_n \to x$ and $y_n \to x$. We claim that x is the least upper bound of E. If $y \in E$, then $y \le x_n$ for all n; let $n \to \infty$ to conclude by Theorem 2.3.5 that $y \le x$, so that x is an upper bound. If x' is an upper bound, then $y_n \le x'$ for all n, and consequently $x \le x'$.

The existence of the greatest lower bound is established by a similar argument that uses lower bounds instead of upper bounds. For example, we let x_1 be a lower bound of E and $y_1 \in E$ so that $x_1 \le y_1$. If m is a lower bound, take $x_2 = m, y_2 = y_1$; if m is not a lower bound, let $y_2 \in E$ with $y_2 < m$, and take $x_2 = x_1$. Inductively, we obtain $C_n = [x_n, y_n]$, where x_n is a lower bound of E and $y_n \in E$. Just as above, if $x \in \bigcap_{n=1}^{\infty} C_n$, then $x = \inf E$. ■

If E is a nonempty subset of R with no upper bound, the smallest element of $\bar{R}$ that is greater than or equal to every member or E is $+\infty$; thus, we may write $\sup E = +\infty$. Similarly if E has no lower bound, $\inf E = -\infty$. Thus, if $+\infty$ and $-\infty$ are allowed, every *nonempty subset of R has a* sup *and an* inf.

The following properties of sups and infs are often used.

2.4.3 THEOREM. *Let E be a nonempty subset of R. If $x < \sup E$, then there is a point $y \in E$ such that $x < y$; if $x > \inf E$, there is a point $y \in E$ such that $x > y$. Furthermore, there is a sequence of points in E converging to* $\sup E$, *and a sequence of points in E converging to* $\inf E$. *Thus, if E is closed, then* $\sup E$ *and* $\inf E$ *(if finite) actually belong to E.*

Proof. Intuitively, if the largest member of a set is greater than x, then some member of the set is greater than x. Formally, if x is smaller than the least upper bound of E, then x cannot be an upper bound. Therefore, x must be smaller than some $y \in E$. (In terms of quantifiers, x is an upper bound of E iff $(\forall y \in E) y \le x$. Thus x is not an upper bound of E iff $(\exists y \in E)(y > x)$.) Similarly, if $x > \inf E$, then x is not a lower bound; hence, $x > y$ for some $y \in E$.

If sup E is finite, then sup $E - 1/n <$ sup E. Hence, by what we have just proved, we can find $x_n \in E$ with sup $E - 1/n < x_n$ (of course $x_n \leq$ sup E). If sup $E = +\infty$, then $n <$ sup E, and thus we can find $x_n \in E$ with $x_n > n$. In either case, $x_n \to$ sup E. To obtain a sequence converging to inf E, use the fact that inf $E + 1/n >$ inf E if inf E is finite, and $-n >$ inf E if inf $E = -\infty$. ■

You may remember the following problem from calculus. You are trying to show that a certain infinite series $\sum_n a_n$ is convergent. If s_n is the nth partial sum, then, for $n > m$,

$$|s_n - s_m| = |a_{m+1} + a_{m+2} + \cdots + a_n| \leq \sum_{j=m+1}^{n} |a_j|.$$

If $\sum_{j=m+1}^{n} |a_j|$ becomes very small for large n and m, we may conclude that $s_n - s_m \to 0$ as $n, m \to \infty$. The inference that s_n must approach a limit is based on a key property of R called completeness, which we now explore.

2.4.4 Definitions and Comments

The sequence $\{x_n\}$ in the metric space Ω is said to be a *Cauchy sequence* if $d(x_n, x_m) \to 0$ as $n, m \to \infty$. In other words, given $\epsilon > 0$ there is a positive integer N such that whenever n and m are greater than or equal to N we have $d(x_n, x_m) < \epsilon$. *Every convergent sequence is Cauchy*, for if $x_n \to x$, then

$$d(x_n, x_m) \leq d(x_n, x) + d(x, x_m) \to 0 \qquad \text{as } n, m \to \infty.$$

Furthermore, *every Cauchy sequence is bounded*, for if x_0 is any point of Ω, $\epsilon > 0$ is arbitrary, and $d(x_n, x_m) < \epsilon$ whenever $n, m \geq N$, then

$$d(x_0, x_n) \leq d(x_0, x_N) + d(x_N, x_n).$$

But $d(x_N, x_n) < \epsilon$ for $n \geq N$, and for $n < N$,

$$d(x_N, x_n) \leq \max_{1 \leq i \leq N-1} d(x_N, x_i).$$

Thus, for some positive constant c, $d(x_0, x_n) \leq c$ for all n, as desired.

A metric space Ω in which every Cauchy sequence converges to an element of Ω is called *complete*. The following result is basic.

2.4.5 THEOREM. *R^p is complete.*

Proof. Let $\{x_n\}$ be a Cauchy sequence in R^p. Since $\{x_n\}$ is bounded, the Bolzano-Weierstrass Theorem 2.3.2 gives us a convergent subsequence $\{x_{n_j}\}$. If $x_{n_j} \to x$, then

$$d(x_n, x) \leq d(x_n, x_{n_j}) + d(x_{n_j}, x).$$

Given $\epsilon > 0$, $\exists N$ such that $d(x_n, x_m) < \epsilon/2$ for $n, m \geq N$. Choose j so large that $n_j \geq N$ and $d(x_{n_j}, x) < \epsilon/2$. If $n \geq N$, it follows that $d(x_n, x) < \epsilon$. Therefore $x_n \to x$. ■

Analysis in R is simplified by the result that *monotone* sequences, that is, sequences that are either increasing ($x_n \leq x_{n+1}$ for all n) or decreasing ($x_n \geq x_{n+1}$) always converge, if we allow $\pm\infty$ as limits.

2.4.6 THEOREM. *If $\{x_n\}$ is a monotone bounded sequence in R, then x_n converges to a limit $x \in R$. An unbounded increasing sequence converges to $+\infty$, and an unbounded decreasing sequence converges to $-\infty$.*

Proof. Assume $x_1 \leq x_2 \leq \ldots$. The decreasing case is handled similarly. Let x be the sup of the set of values $\{x_1, x_2, \ldots\}$. In the bounded case, x is finite. If $\epsilon > 0$, then $x - \epsilon < x$, so by Theorem 2.4.3, $x - \epsilon < x_{n_0}$ $(\leq x)$ for some n_0. By monotonicity, $x - \epsilon < x_n \leq x$ for all $n \geq n_0$, and consequently $x_n \to x$. In the unbounded case, $x = +\infty$, and again by Theorem 2.4.3 for any positive number M we can find $x_{n_0} > M$, so $x_n > M$ for all $n \geq n_0$. Thus, $x_n \to \infty$. ■

Problems for Section 2.4

1. In Theorem 2.4.2, obtain the existence of the greatest lower bound by an alternative approach, namely, by considering $-E = \{-x : x \in E\}$.
2. If $\{x_n\}$ is a Cauchy sequence with a subsequence converging to x, show that $x_n \to x$.

3. Let K be a compact subset of the metric space Ω. Show that K is complete; that is, any Cauchy sequence in K converges to a point of K.
4. Give an example of a metric space that is not complete.
5. If K is a subset of R^p such that every sequence of points in K has a subsequence converging to a point of K, show that K is closed and bounded, and hence compact.
6. Define $x_{n+1} = \sqrt{2 + \sqrt{x_n}}$, $n = 1, 2, \ldots$, with $x_1 = \sqrt{2}$.
   (a) Show that $\{x_n\}$ is a monotone bounded sequence and, hence, converges to a finite limit L.
   (b) Find L (approximately).
7. Let V be an open subset of R. If $x \in V$, let V_x be the union of all open intervals I such that $x \in I$ and $I \subseteq V$ (see Section 2.1, Problems 3 and 4). Define

$$b_x = \sup\{y : (x, y) \subseteq V\},$$

$$a_x = \inf\{z : (z, x) \subseteq V\},$$

   and show that $V_x = (a_x, b_x)$.
8. Continuing Problem 7, show that every open subset of R can be expressed as a disjoint union of countably many open intervals.
9. True or false: if E is a nonempty subset of R, then sup E is a limit point of E.
10. True or false: if sup E and inf E belong to E then E is closed.

REVIEW PROBLEMS FOR CHAPTER 2

1. Give an example of a closed set of real numbers that is not compact.
2. Let E be a nonempty subset of R, and assume that E has a least upper bound $x \in R$. If x does not belong to E, show that x is a limit point of E.
3. Recall (Section 1.5, Problem 8) that a point x is a *boundary point* of the set E if every open ball $B_r(x)$ contains both a point of E and a point of E^c. If E is a nonempty, bounded subset of R, show that E has at least one boundary point.

4. A sequence is defined inductively by

$$x_1 = 1, \quad x_{n+1} = \frac{12}{1 + x_n}, \qquad n \geq 1.$$

If $\{x_n\}$ is known to converge to a limit L, find L explicitly.

5. Let $x_1 > 1, x_{n+1} = 2 - (1/x_n), n \geq 1$. Show that $\{x_n\}$ converges, and find the limit.

6. Let f and g be bounded functions from Ω to R; in other words, for some $M > 0$ we have $|f(x)| \leq M$ for all $x \in \Omega$, and similarly for g. Show that

$$\sup_{x \in \Omega} [f(x) + g(x)] \leq \sup_{x \in \Omega} f(x) + \sup_{x \in \Omega} g(x),$$

where, for example, $\sup_{x \in \Omega} f(x)$ is the least upper bound of the set of all $f(x)$, $x \in \Omega$. (If you have trouble, think intuitively about maximizing $f(x) + g(x)$ as x ranges over Ω.)

3

UPPER AND LOWER LIMITS OF SEQUENCES OF REAL NUMBERS

3.1 GENERALIZATION OF THE LIMIT CONCEPT

In calculus it was often necessary to compute the limit of a particular sequence, and occasionally a sequence having no limit at all was encountered. For example, consider the sequence

$$\frac{1}{2}, \frac{1}{2}, -\frac{1}{2}, \frac{2}{3}, \frac{1}{3}, -\frac{3}{4}, \frac{3}{4}, \frac{1}{4}, -\frac{7}{8}, \frac{4}{5}, \frac{1}{5}, -\frac{15}{16}, \ldots .$$

The sequence $\{x_n\}$ has no limit, but the subsequence $x_1, x_4, x_7, \ldots$ converges to 1, the subsequence $x_2, x_5, x_8, \ldots$ converges to 0, and the subsequence $x_3, x_6, x_9, \ldots$ converges to -1. When we encounter sequences having no limit, it will be useful for us to work with the largest of all limits of subsequences ($+1$ in the above example) and the smallest of all subsequential limits (-1 above).

3.1.1 Definitions and Comments

Let $\{x_n\}$ be a sequence of real numbers; by Theorem 2.3.4, $\{x_n\}$ has at least one convergent subsequence, if $+\infty$ and $-\infty$ are allowed as limits. Thus the set S of all subsequential limits is a nonempty subset of the extended reals $\bar{R}$. It follows that S has a sup and an inf. For clearly ∞ is an upper bound of S; if ∞ is not the least upper bound, then S has a real upper bound, and consequently S has a sup by Theorem 2.4.2. The argument that S has an inf is similar.

We define the *upper limit* of the sequence $\{x_n\}$ as sup S and the *lower limit* of $\{x_n\}$ as inf S.

The standard notation for the upper limit is

$$\limsup_{n\to\infty} x_n \qquad \text{or simply} \qquad \limsup x_n,$$

and the lower limit is denoted by

$$\liminf_{n\to\infty} x_n \qquad \text{or simply} \qquad \liminf x_n.$$

In fact we can show that there are subsequences converging to sup S and inf S, and therefore sup S and inf S actually belong to S. Thus, lim sup x_n is the largest element of S, in other words, the *largest subsequential limit*, and lim inf x_n is the smallest element of S, the *smallest subsequential limit*.

To see this, let $s = \sup S$; we must show that $s \in S$. We know that there is a sequence of points $s_1, s_2, \ldots$ in S such that $s_k \to s$ as $k \to \infty$ (see Theorem 2.4.3). Since $s_k \in S$ we have a subsequence $x_{k1}, x_{k2}, \ldots$ of $\{x_n\}$ converging to s_k. We summarize the relevant information as follows:

$$\begin{array}{ccccc}
x_{11} & x_{12} & x_{13} & \ldots \to & s_1 \\
x_{21} & x_{22} & x_{23} & \ldots \to & s_2 \\
 & \vdots & & & \\
x_{n1} & x_{n2} & x_{n3} & \ldots \to & s_n \\
 & \vdots & & &
\end{array}$$

First assume that s is finite. For each n pick $m = m_n$ such that $|x_{nj} - s_n| < 1/n$ for all $j \geq m_n$. Then consider the "diagonal sequence" (a useful description, although we do not necessarily go down the diagonal of the above array) $x_{1m_1}, x_{2m_2}, x_{3m_3}, \ldots$. We adjust the indices m_n so as to obtain a true subsequence; in other words, if, for example, $x_{1m_1} = x_7, x_{2m_2} = x_6$, we increase m_2 so that x_{2m_2} becomes x_r for some $r > 7$. Now

$$|x_{nm_n} - s_n| < 1/n \qquad \text{and} \qquad s_n \to s;$$

hence,

$$|x_{nm_n} - s| \le |x_{nm_n} - s_n| + |s_n - s| \to 0.$$

Therefore, $x_{nm_n} \to s$, so $s \in S$. Since $s = \sup S$, it follows that s is the largest element of S.

Now assume $s = \infty$. (If $s = -\infty$, then S must consist of $-\infty$ alone; therefore $s \in S$. Alternatively, a direct argument similar to the $s = \infty$ case can be made). If $\infty \in S$ we are finished, so assume $\infty \notin S$. As in Theorem 2.4.3, we can obtain a sequence of finite numbers $s_n \in S$ with $s_n \to \infty$. If the subsequence $\{x_{nm_n}\}$ is constructed as above, then $x_{nm_n} > s_n - 1/n$, so that $x_{nm_n} \to \infty$. Thus $\infty \in S$, as desired.

The above argument shows that if $s_1, s_2, \ldots \in S$ and $s_n \to t$ then $t \in S$; thus *S is a closed set in $\bar{R}$*.

Upper and lower limits form a natural generalization of the basic limit concept, as the next result shows.

3.1.2 THEOREM. *Suppose $\{x_n\}$ is a sequence of real numbers. If $\lim_{n\to\infty} x_n = x$, then $\limsup x_n = \liminf x_n = x$. Conversely, if $\limsup x_n = \liminf x_n = x$, then $x_n \to x$.*

Proof. If $x_n \to x$, then all subsequences converge to x also, and hence $S = \{x\}$. Therefore, $\limsup x_n = \liminf x_n = x$. Conversely, suppose x_n does not converge to x. Assume first that x is finite. The statement $x_n \to x$ means that for every $\epsilon > 0$ there is a positive integer N such that for all $n \ge N$ we have $|x_n - x| < \epsilon$. Symbolically,

$$(\forall \epsilon > 0)(\exists N)(\forall n \ge N)(|x_n - x| < \epsilon).$$

We know how to find the negation of statement of this type (see Section 1.4.5).

The statement $x_n \not\to x$ means

$$(\exists\epsilon > 0)(\forall N)(\exists n \geq N)(|x_n - x| \geq \epsilon).$$

In other words, there is an $\epsilon > 0$ such that for all N we can find an $n \geq N$ with $|x_n - x| \geq \epsilon$. Set $N = 1$ and find $n = n_1$ such that $|x_{n_1} - x| \geq \epsilon$. Then set $N = n_1 + 1$ and find $n = n_2 \geq N$ such that $|x_{n_2} - x| \geq \epsilon$. Proceeding in this fashion, we obtain a sequence $x_{n_1}, x_{n_2}, \ldots$ with $|x_{n_j} - x| \geq \epsilon$ for all j. This sequence has a convergent subsequence by Theorem 2.3.4, and the limit must be outside the interval $(x - \epsilon, x + \epsilon)$. But if $\limsup x_n = \liminf x_n = x$, we must have $S = \{x\}$, a contradiction.

Now if $x = \infty$, we write the definition of $x_n \to \infty$:

$$(\forall M > 0)(\exists N)(\forall n \geq N)(x_n > M);$$

thus, $x_n \not\to \infty$ means

$$(\exists M > 0)(\forall N)(\exists n \geq N)(x_n \leq M).$$

In other words, there is a positive real number M such that for every positive integer N we can find $n \geq N$ with $x_n \leq M$. As above, we find a subsequence with a limit that is less than or equal to M, contradicting the fact that $S = \{\infty\}$. If $x = -\infty$, the argument is similar. ■

Problems for Section 3.1

1. Let $S = \{-3, 2, 4, 10\}$. Construct a sequence $\{x_n\}$ of real numbers such that the set of subsequential limits of $\{x_n\}$ is precisely S.
2. Suppose $x_{n+1} = f(n)x_n$, where $f(n) \to 3$ as $n \to \infty$. What are the possible values of $\limsup x_n$ and $\liminf x_n$? Does $\{x_n\}$ always converge?
3. Let c be a positive real number. Show that $\liminf(cx_n) = c \liminf x_n$ and $\limsup(cx_n) = c \limsup x_n$.
4. Let c be a negative real number. Show that $\liminf(cx_n) = c \limsup x_n$ and $\limsup(cx_n) = c \liminf x_n$.

3.2 SOME PROPERTIES OF UPPER AND LOWER LIMITS

Before looking at further properties of upper and lower limits, we need some additional terminology.

3.2.1 Definitions and Comments

Let $P_1, P_2, \ldots$ be a sequence of statements (for example, P_n might be "$x_n < 3$," where $\{x_n\}$ is a sequence of real numbers). We say that P_n holds *eventually* (sometimes abreviated *ev.*) if P_n holds for all but finitely many n. For example, if P_n holds for all $n \geq 100$, then P_n holds ev. For $n < 100$, P_n might be true for some n and false for others, but it doesn't matter; once we get to $n = 100$, P_n is true from that point on. We say that P_n holds *infinitely often* (sometimes abbreviated $i.o.$) if P_n is true for infinitely many n. For example, if P_n is true whenever n is even, then P_n holds i.o., regardless of what happens when n is odd. Note that if P_n holds eventually, then P_n holds infinitely often, but not conversely.

If P_n^c is the negation of P_n, then the negation of "P_n holds infinitely often" is "P_n holds for only finitely many n"; that is, "P_n^c holds for all but finitely many n"; in other words, "P_n^c holds eventually." If we replace P_n by P_n^c, we see that the negation of "P_n holds eventually" is "P_n^c holds infinitely often." These ideas will now be used in establishing basic properties of upper and lower limits.

3.2.2 THEOREM. *Let $\{x_n\}$ be a sequence of real numbers, and assume that*
lim sup $x_n = x$, lim inf $x_n = y$.

(a) *If $z > x$, then $x_n < z$ eventually.*
(b) *If $z < x$, then $x_n > z$ infinitely often.*

Furthermore, (a) *and* (b) *characterize the* lim sup; *that is, if x' is a number in $\bar{R}$ satisfying* (a) *and* (b), *then $x' = x$. Similarly:*

(c) *If $z < y$, then $x_n > z$ eventually.*
(d) *If $z > y$, then $x_n < z$ infinitely often.*

If y' is a number in $\bar{R}$ satisfying (c) *and* (d), *then* $y' = y$.

Proof

(a) If "$x_n < z$ ev." is false, then by Section 3.2.1, $x_n \geq z$ i.o., and, hence, by Theorem 2.3.5 there is a subsequence coverging to a limit that is at least z. Thus, $x = \limsup x_n \geq z$.

(b) If "$x_n > z$ i.o." is false, then by Section 3.2.1, $x_n \leq z$ ev., and therefore all subsequential limits are less than or equal to z. Thus, $x = \limsup x_n \leq z$. (Alternatively, there is a subsequence $x_{n_k} \to x$ so that if $z < x$ then, for all sufficiently large k, $x_{n_k} > z$.)

(c) If "$x_n > z$ ev." is false, then $x_n \leq z$ i.o., and it follows as in (a) that $y \leq z$.

(d) If "$x_n < z$ i.o." is false, then $x_n \geq z$ ev., and therefore, as in (b), $y \geq z$. (Alternatively, there is a subsequence $x_{n_k} \to y$ so that if $z > y$ then $x_{n_k} < z$ for all sufficiently large k.)

Now suppose x' satisfies (a) and (b). Assume (without loss of generality) that $x < t < x'$. Since $t < x'$, (b) gives $x_n > t$ i.o.; since $t > x$, (a) gives $x_n < t$ ev. This is a contradiction. Similarly, if y' satisfies (c) and (d) and $y < u < y'$, then by (c), $x_n > u$ ev., and by (d), $x_n < u$ i.o., a contradiction. ■

The last part of the proof is a typical "uniqueness argument."

Theorem 3.2.2 may be used to show that $\limsup x_n$ is really the "limit of a sup," and similarly for $\liminf x_n$.

3.2.3 COROLLARY

$$\limsup_{n\to\infty} x_n = \lim_{n\to\infty} \left(\sup_{k\geq n} x_k\right);$$

$$\liminf_{n\to\infty} x_n = \lim_{n\to\infty} \left(\inf_{k\geq n} x_k\right).$$

Proof. As n increases, we are taking the sup of a smaller set in evaluating $\sup_{k\geq n} x_k$, so by Theorem 2.4.6, $\sup_{k\geq n} x_k$ decreases to a limit x. (For example, if $\{x_n\} = \frac{1}{2}, -\frac{1}{2}, \frac{1}{3}, -\frac{2}{3}, \frac{1}{4}, -\frac{3}{4}, \ldots,$ then

$$\sup_{k\geq 1} x_k = \frac{1}{2}; \quad \sup_{k\geq 2} x_k = \frac{1}{3}; \quad \sup_{k\geq 3} x_k = \frac{1}{3}, \text{ etc.})$$

Similarly, $\inf_{k\geq n} x_k$ will increase to a limit y. If $z > x$, then eventually, $\sup_{k\geq n} x_k < z$, so $x_k < z$ for all $k \geq n$. Thus, $x_n < z$ ev. If $z < x$, then for all n, $\sup_{k\geq n} x_k > z$, so for all n there is, by Theorem 2.4.3, a positive integer $k \geq n$ such that $x_k > z$. In other words, $x_n > z$ i.o. By Theorem 3.2.2, x must be lim sup x_n.

If $z < y$, then $\inf_{k\geq n} x_k > z$ ev., so $x_k > z$ for all $k \geq n$. Thus, $x_n > z$ ev. If $z > y$, then $\inf_{k\geq n} x_k < z$ for all n, so for all n there is, by Theorem 2.4.3, a positive integer $k \geq n$ such that $x_n < z$. Thus, $x_n < z$ i.o. By Theorem 3.2.2, y must be lim inf x_n. ■

3.2.4 Remark

In proving Corollary 3.2.3, we saw that $\sup_{k\leq n} x_k$ decreases and $\inf_{k\geq n} x_k$ increases as n increases. Thus, we may write

$$\limsup_{n\to\infty} x_n = \inf_n \sup_{k\geq n} x_k,$$

$$\liminf_{n\to\infty} x_n = \sup_n \inf_{k\geq n} x_k.$$

Problems for Section 3.2

1. Let $\{x_n\}$ and $\{y_n\}$ be sequences of real numbers. Show that if $x_n \leq y_n$ for all n then lim inf $x_n \leq$ lim inf y_n and lim sup $x_n \leq$ lim sup y_n.
2. Show that lim sup$(x_n + y_n) \leq$ lim sup x_n + lim sup y_n and

$$\liminf(x_n + y_n) \geq \liminf x_n + \liminf y_n$$

(assuming in each case that the right-hand side is not of the form $+\infty - \infty$ or $-\infty + \infty$).

In Section 3.3, we study convergence of power series, so it will be useful to look at a few problems involving finite and infinite series of real numbers.

3. Let $a_1, \ldots, a_n$, $b_1, \ldots, b_n$ be real numbers. Prove the *Cauchy-Schwarz inequality*

$$\left(\sum_{j=1}^{n} a_j b_j\right)^2 \leq \left(\sum_{j=1}^{n} a_j^2\right)\left(\sum_{j=1}^{n} b_j^2\right).$$

[Hint:

$$\sum_{j=1}^{n} (a_j x + b_j)^2 \geq 0 \qquad \text{for all real } x.]$$

4. Use Problem 3 to show that if $a_n \geq 0$ for all n and $\sum_{n=1}^{\infty} a_n$ converges, then $\sum_{n=1}^{\infty} \sqrt{a_n}/n$ converges. Can you obtain this result without using Problem 3?
5. In R^p, define $|x| = (\sum_{i=1}^{n} x_i^2)^{1/2}$, where $x = (x_1, \ldots, x_n)$. Then the Euclidean metric is given by $d(x, y) = |x - y|$. Use Problem 3 to show that d is actually a metric.
6. Express the statement "$x_n > 3$ eventually" using existential and universal quantifiers. Then use the technique of Section 1.4.5 to take the negation and verify that the result is "$x_n \leq 3$ infinitely often."

3.3 CONVERGENCE OF POWER SERIES

An important application of upper and lower limits is to the problem of convergence of power series. (A *power series* is an expression of the form $\sum_{n=0}^{\infty} c_n (z - z_0)^n$.) The natural domain for the discussion is the complex plane C rather than the set R of real numbers. One reason is that, given any differentiable function f defined on an open subset V of C and a point $z_0 \in V$, there is a power series representation $f(z) = \sum_{n=0}^{\infty} c_n (z - z_0)^n$ valid in a neighborhood of z_0. The corresponding result for real-valued functions is false. However, no knowledge of complex variable theory is needed here; we only use the absolute value (magnitude) of a complex number; that is, $|a + ib| = (a^2 + b^2)^{1/2}$. You can assume if you like that all numbers in the discussion to follow are real, and no harm will be done. Also, for convenience, we assume $z_0 = 0$; this amounts to making a change of variable $w = z - z_0$, so no generality is lost.

Let's begin with an example. Suppose we are given the power series $\sum_{n=0}^{\infty}(\frac{1}{2})^n z^n$; for which values of z does the series converge? (Converge of the series means convergence of the sequence of partial sums $s_n = \sum_{k=0}^{n}(\frac{1}{2})^k z^k$. Also, in the discussion to follow, "convergence" always means "convergence to a finite limit".) A technique from calculus called the *ratio test* works in most "practical" examples. We find the limit of the absolute value of the ratio of the $(n+1)$st term of the series to the nth term:

$$\left|\frac{(\frac{1}{2}z)^{n+1}}{(\frac{1}{2}z)^n}\right| = \frac{1}{2}|z| \to \frac{1}{2}|z| \quad \text{as} \quad n \to \infty.$$

If the limit is less than 1, that is, $|z| < 2$, the series converges. If the limit is greater than 1, that is, $|z| > 2$, the series diverges. If the limit is 1, that is, $|z| = 2$, the test gives no information.

In general, the limit involved in the ratio test might not exist, so a different test is used.

3.3.1 Theorem (Root Test)

Let $\{a_n\}$ be a sequence of real numbers, and define

$$a = \limsup_{n\to\infty}(|a_n|)^{1/n}.$$

If $a < 1$, the series $\sum a_n$ converges; in fact, it converges absolutely (that is, the series $\sum |a_n|$ converges). If $a > 1$, the series diverges; if $a = 1$, the test gives no information.

Proof. If $a < 1$, pick b such that $a < b < 1$. By Theorem 3.2.2(a), $|a_n|^{1/n} < b$ eventually, so $|a_n| < b^n$ eventually. Thus, $\sum |a_n|$ converges by comparison with a geometric series. Now absolute convergence implies convergence, for if $s_n = \sum_{k=1}^{n} a_k$ then $|s_n - s_m| = |\sum_{k=m+1}^{n} a_k| \le \sum_{k=m+1}^{n} |a_k| \to 0$ as $n, m \to \infty$. (Recall from calculus that convergence does not imply absolute convergence; the standard example is $1 - \frac{1}{2} + \frac{1}{3} - \frac{1}{4} + \cdots$).

If $a > 1$, pick b such that $a > b > 1$. By Theorem 3.2.2(b), $|a_n|^{1/n} > b$ infinitely often, or $|a_n| > b^n$ infinitely often; consequently, a_n cannot approach 0 as $n \to \infty$. The series therefore diverges, for if $\sum_n a_n$ converges to s, then $a_n = \sum_{k=1}^{n} a_k - \sum_{k=1}^{n-1} a_k \to s - s = 0$.

To show that $a = 1$ provides no information, consider the divergent series $\sum 1/n$ and the convergent series $\sum 1/n^2$. Since $(1/n^r)^{1/n} \to 1$ as $n \to \infty$ for any $r > 0$ (take logarithms to verify this), $a = 1$ in both cases. (The root test works for complex numbers also; the proof is the same.) ■

Now consider the power series $\sum_{n=0}^{\infty} c_n z^n$. We look at the series of absolute values $\sum_{n=0}^{\infty} |c_n||z|^n$, which brings the problem back to real variables. We have the following result.

3.3.2 THEOREM. *Let $a_0 = \limsup |c_n|^{1/n}$, and let $r = 1/a_0$ (take $r = \infty$ if $a_0 = 0$). If $|z| < r$, the power series $\sum c_n z^n$ converges absolutely, and if $|z| > r$ the series diverges.*

Proof. We apply the root test to $\sum |c_n||z|^n$; we compute

$$\limsup(|c_n||z|^n)^{1/n} = |z| \limsup |c_n|^{1/n} = |z| a_0.$$

The result follows from Theorem 3.3.1. ■

The number r defined in Theorem 3.3.2 is called the *radius of convergence* of the series. The series converges inside the circle of radius r and center at 0 (in the real case, we have instead an interval of length r) and diverges outside the circle. The case $r = \infty$ is not uncommon; for example, the familiar expansions $e^z = \sum_{n=0}^{\infty} z^n/n!$, $\cos z = 1 - z^2/2! + z^4/4! - \dots$, and $\sin z = z - z^3/3! + z^5/5! - \dots$ converge for all z.

Problems for Section 3.3

1. Consider the series

$$\frac{1}{2} + \frac{1}{3} + \frac{1}{4} + \frac{1}{9} + \frac{1}{8} + \frac{1}{27} + \frac{1}{16} + \cdots$$
$$+ \frac{1}{2^k} + \frac{1}{3^k} + \frac{1}{2^{k+1}} + \frac{1}{3^{k+1}} + \cdots.$$

Verify that the ratio test cannot be applied, but the root test yields convergence.

2. Find the radius of convergence of each of the following power series:

   (a) $\sum n^k z^n$, k a positive integer

   (b) $\sum \dfrac{3^n}{n^5} z^n$

3. Suppose the power series coefficients c_n have the property that $|c_n|^{1/n} \geq 3$ infinitely often. What can be said about the radius of convergence of the series?

4. A series that is convergent but not absolutely convergent can be rearranged so as not to converge to the same sum. For example, if

$$S = 1 - \frac{1}{2} + \frac{1}{3} - \frac{1}{4} + \frac{1}{5} - \frac{1}{6} + \ldots,$$

   show that

$$1 + \frac{1}{3} - \frac{1}{2} + \frac{1}{5} + \frac{1}{7} - \frac{1}{4} + \frac{1}{9} + \frac{1}{11} - \frac{1}{6} + \cdots = \frac{3}{2}S.$$

5. Continuing Problem 4, a conditionally but not absolutely convergent series can be rearranged to converge to any given real number r, or to diverge to $+\infty$, or to diverge to $-\infty$, or simply to diverge (not to $\pm\infty$). Can you describe a strategy for accomplishing this?

6. Suppose $\sum_{n=0}^{\infty} c_n x^n$ converges to $f(x)$ for $-1 < x < 1$, and let $s_n = c_0 + \cdots + c_n$, $s_{-1} = 0$. Assume $s_n \to s$ (finite) as $n \to \infty$.

   (a) Show that $f(x) = (1 - x) \sum_{n=0}^{\infty} s_n x^n$ for $|x| < 1$.

   (b) Given $\epsilon > 0$, choose N so that $|s - s_n| < \epsilon/2$ for all $n > N$. Show that

$$|f(x) - s| \leq (1 - x) \sum_{n=0}^{N} |s_n - s| x^n + \frac{\epsilon}{2}, \qquad 0 < x < 1.$$

   (c) Prove *Abel's Theorem*: If $f(x) = \sum_{n=0}^{\infty} c_n x^n$, $-r < x < r$, where $\sum_{n=0}^{\infty} c_n r^n$ converges, then

$$\lim_{x \to r} f(x) = \sum_{n=0}^{\infty} c_n r^n.$$

REVIEW PROBLEMS FOR CHAPTER 3

1. Give an example of
   (a) a sequence of real numbers such that $\limsup x_n \neq \liminf x_n$.
   (b) a sequence $\{x_n\}$ of real numbers and an open interval I such that $x_n \in I$ infinitely often, but x_n is not eventually in I.
2. Give an example of a power series whose radius of convergence is 2.
3. True or false: the radius of convergence of a power series is always strictly greater than zero.
4. True or false, and explain briefly ($\{x_n\}$ is a sequence of real numbers throughout).
   (a) If $\lim\ \sup\ x_n = 5$, then $x_n > 4$ eventually.
   (b) If $\lim\ \sup\ x_n = 5$, then $x_n > 4$ infinitely often.
   (c) If $\lim\ \inf\ x_n = 5$, then $x_n > 4$ eventually.
   (d) If $x_n > 4$ eventually, then $\lim\ \inf\ x_n > 4$.
5. Give an example of a sequence of real numbers whose set of subsequential limits is $S = \{-\infty, \infty\}$.
6. Let

$$A_n = \begin{cases} [0,1] & \text{if } n = 1,3,5,7,\ldots, \\ [-1,0] & \text{if } n = 2,4,6,8,\ldots. \end{cases}$$

Let A be the set of all real numbers x such that $x \in A_n$ infinitely often, and let B be the set of real numbers x such that $x \in A_n$ eventually. Find A and B.

4

CONTINUOUS FUNCTIONS

4.1 CONTINUITY: IDEAS, BASIC TERMINOLOGY, PROPERTIES

The idea of continuity of a function f at a point x is familiar. Intuitively, as y gets close to x, $f(y)$ gets close to $f(x)$. This has a direct translation in terms of sequences.

4.1.1 Definition

Let f be a function defined on the metric space Ω and taking values in the metric space Ω'; the standard notation is $f\colon \Omega \to \Omega'$. The set Ω is called the *domain* of f, and Ω' the *codomain* of f; f is sometimes called a *mapping*. (If it aids the understanding, you may assume $\Omega = \Omega' = R$.) If $x \in \Omega$, we say that f is *continuous at* x if for all sequences $x_1, x_2, \ldots$ in Ω with $x_n \to x$, we have $f(x_n) \to f(x)$.

There is another way of describing continuity, the so-called epsilon-delta approach.

4.1.2 THEOREM. *The function f is continuous at x if and only if for every $\epsilon > 0$ there is a $\delta > 0$ such that whenever*[1] *y is a point of Ω with $d(x,y) < \delta$, we have $d(f(x), f(y)) < \epsilon$.*

Thus, if we wish to force $f(y)$ to be "close" to $f(x)$, that is, within distance ϵ, we can accomplish this by taking y "close" to x, in other words, within distance δ. (We have followed common practice and used the same letter d for the metrics on Ω and Ω'. This should cause no confusion since in each expression involving d it is clear which space is being considered).

Proof. We may write the epsilon-delta condition described in Theorem 4.1.2 as follows (the symbol $\Rightarrow$ denotes "implies"):

$$(1)\quad (\forall\epsilon > 0)(\exists\delta > 0)(\forall y \in \Omega)(d(x,y) < \delta \Rightarrow d(f(x), f(y)) < \epsilon).$$

The negation of this statement is

$$(2)\quad (\exists\epsilon > 0)(\forall\delta > 0)(\exists y \in \Omega)(d(x,y) < \delta \text{ but } d(f(x), f(y)) \geq \epsilon).$$

Thus, if (1) is false then we can find some $\epsilon > 0$ such that no matter which $\delta > 0$ is selected, there is a point y within distance δ of x such that $f(y)$ is at least distance ϵ from $f(x)$. Since any choice of δ is allowed, we can take $\delta = 1/n$, $n = 1, 2, \ldots$; label the corresponding point y as x_n. Then $d(x, x_n) < 1/n$, and consequently $x_n \to x$. But $d(f(x), f(x_n)) \geq \epsilon$ for all n, so $f(x_n) \not\to f(x)$. We conclude that f is discontinuous at x.

Now assume that (1) holds, and let $x_n \to x$. Given $\epsilon > 0$, choose $\delta > 0$ so that (1) is satisfied. Since $x_n \to x$, we have $d(x, x_n) < \delta$ eventually, so by (1), $d(f(x), f(x_n)) < \epsilon$ eventually. It follows that $f(x_n) \to f(x)$. ■

It is possible to show directly that a familiar function such as a polynomial is continuous (see Problems 1–3), but the easiest way to do this is to show that the function is differentiable. In the next chapter we study differentiation, and show that differentiability implies continuity.

[1] The "whenever" phrase is translated as "for every y in Ω such that $d(x,y) < \delta$," etc.

A basic continuity result involves composition of functions, which we now describe. If $f:\Omega \to \Omega'$ and $g:\Omega' \to \Omega''$, we define the *composition* of f and g as the function $h:\Omega \to \Omega''$ given by

$$h(x) = g(f(x)), \qquad x \in \Omega.$$

Thus, given x we compute $f(x) \in \Omega'$ and then apply g to $f(x)$ to get $h(x)$ (see Fig. 4.1.1). The standard notation for composition is $g \circ f$, sometimes read "f followed by g." ■

4.1.3 THEOREM. *The composition of continuous functions is continuous; that is, if f is continuous at x and g is continuous at $f(x)$, then $g \circ f$ is continuous at x.*

Proof. If $x_n \to x$, then $f(x_n) \to f(x)$; hence, $g(f(x_n)) \to g(f(x))$. ■

We have described continuity at a particular point of Ω; "continuity on Ω" simply means continuity at each point. However, the *global* concept of continuity on the entire space (as opposed to the *local* idea of continuity at a single point) has an important description in terms of open and closed sets. Before discussing this, one more bit of terminology is needed.

4.1.4 Definition

If $f:\Omega \to \Omega'$ and A is a subset of Ω', the *preimage* (or *inverse image*) of A under f is

$$f^{-1}(A) = \{x \in \Omega : f(x) \in A\}.$$

In other words,

$$x \in f^{-1}(A) \quad \text{iff} \quad f(x) \in A.$$

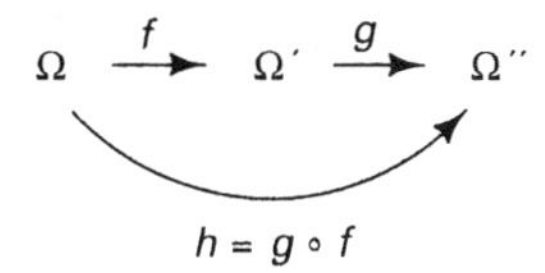

Figure 4.1.1 Composition of Functions

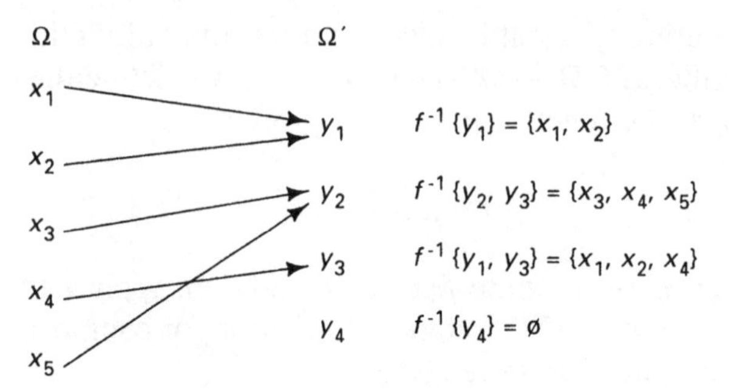

Figure 4.1.2 Preimages

Thus $f^{-1}(A)$ is the set of points mapped by f into A. See Fig. 4.1.2 for an explicit calculation in a simple case.

Preimages behave very well with respect to unions, intersections, and complements, as the following result shows.

4.1.5 THEOREM. *Let $f:\Omega \to \Omega'$, and let $\{A_i\}$ be an arbitrary family (there may be uncountably many i) of subsets of Ω'. Then*

$$f^{-1}\left(\bigcup_i A_i\right) = \bigcup_i f^{-1}(A_i), \qquad f^{-1}\left(\bigcap_i A_i\right) = \bigcap_i f^{-1}(A_i),$$

and

$$f^{-1}(A_i^c) = [f^{-1}(A_i)]^c.$$

Also, if $A_1 \subseteq A_2$, then

$$f^{-1}(A_1) \subseteq f^{-1}(A_2).$$

Proof

$$\begin{aligned} x \in f^{-1}\left(\bigcup_i A_i\right) \quad &\text{iff} \quad f(x) \in \bigcup_i A_i \\ &\text{iff} \quad f(x) \in A_i, \quad \text{that is,} \quad x \in f^{-1}(A_i), \\ &\qquad \text{for at least one} \quad i \\ &\text{iff} \quad x \in \bigcup_i f^{-1}(A_i); \end{aligned}$$

$$
\begin{aligned}
x \in f^{-1}\left(\bigcap_i A_i\right) \quad &\text{iff} \quad f(x) \in \bigcap_i A_i \\
&\text{iff} \quad f(x) \in A_i, \quad \text{that is,} \quad x \in f^{-1}(A_i), \\
&\qquad \text{for every} \quad i \\
&\text{iff} \quad x \in \bigcap_i f^{-1}(A_i); \\
x \in f^{-1}(A_i^c) \quad &\text{iff} \quad f(x) \in A_i^c \quad \text{iff} \quad f(x) \notin A_i \\
&\text{iff} \quad x \notin f^{-1}(A_i) \quad \text{iff} \quad x \in [f^{-1}A_i]^c.
\end{aligned}
$$

The final statement is a consequence of the definition of preimage. ■

We may now relate continuity to open and closed sets.

4.1.6 THEOREM. *Let $f:\Omega \to \Omega'$, where Ω and Ω' are metric spaces. The function f is continuous on Ω if and only if for each open set $V \subseteq \Omega'$ the preimage $f^{-1}(V)$ is an open subset of Ω. Equivalently, for each closed set $C \subseteq \Omega'$, the preimage $f^{-1}(C)$ is a closed subset of Ω.*

Proof. Assume f continuous. Let x belong to $f^{-1}(V)$, where V is open in Ω'. Then $f(x) \in V$, so for some $\epsilon > 0$, $B_\epsilon(f(x)) \subseteq V$. If $\delta > 0$ is as given by the statement that f is continuous at x (see Theorem 4.1.2), then

$$y \in B_\delta(x) \Rightarrow f(y) \in B_\epsilon(f(x)); \quad \text{hence, } f(y) \in V.$$

Thus, $y \in f^{-1}(V)$, proving that $B_\delta(x) \subseteq f^{-1}(V)$. Therefore $f^{-1}(V)$ is open.

Conversely, assume V open implies $f^{-1}(V)$ open. If $x \in \Omega$, we show that f is continuous at x. Given $\epsilon > 0$, $f(x) \in B_\epsilon(f(x))$, which is an open set V. Thus, $x \in f^{-1}(V)$, which is open by hypothesis, so $B_\delta(x) \subseteq f^{-1}(V)$ for some $\delta > 0$. Consequently,

$$y \in B_\delta(x) \Rightarrow y \in f^{-1}(V) \Rightarrow f(y) \in V = B_\epsilon(f(x));$$

in other words,

$$d(x,y) < \delta \Rightarrow d(f(x), f(y)) < \epsilon.$$

By Theorem 4.1.2, f is continuous at x.

Finally, we must show that the preimage of each closed set is closed if and only if the preimage of each open set is open. Suppose that for each closed $C \subseteq \Omega'$, $f^{-1}(C)$ is closed, and assume V is an open subset of Ω'. Then V^c is closed, so $f^{-1}(V^c)$ is closed. But by Theorem 4.1.5,

$$f^{-1}(V^c) = [f^{-1}(V)]^c.$$

Thus, $[f^{-1}(V)]^c$ is closed, so $f^{-1}(V)$ is open. Conversely, if the preimage of each open set is open and C is a closed subset of Ω', then C^c is open, and hence $f^{-1}(C^c) = [f^{-1}(C)]^c$ is open. Therefore $f^{-1}(C)$ is closed. ■

A companion to the preimage idea is that of the direct image, defined as follows.

4.1.7 Definition

If $f: \Omega \to \Omega'$ and A is a subset of Ω, the *image* (or *direct image*) of A under f is

$$f(A) = \{f(x) : x \in A\};$$

thus, $f(A)$ is the set of all values $f(x)$ as x ranges over A.

Direct images are not as well behaved as preimages (see Problem 4), but they have some desirable properties.

4.1.8 THEOREM. *Let $f: \Omega \to \Omega'$. Then*

(a) If $A \subseteq \Omega$, then $A \subseteq f^{-1}[f(A)]$.
(b) If $B \subseteq \Omega'$, then $f[f^{-1}(B)] \subseteq B$.
(c) If $A \subseteq C \subseteq \Omega$, then $f(A) \subseteq f(C)$.
(d) If $\{A_i\}$ is an arbitrary family of subsets of Ω, then

$$f\left(\bigcup_i A_i\right) = \bigcup_i f(A_i).$$

Proof

(a) If $x \in A$, then $f(x) \in f(A)$; hence, $x \in f^{-1}[f(A)]$. Note that A can be a proper subset of $f^{-1}[f(A)]$; see Fig. 4.1.3.
(b) If $y \in f[f^{-1}(B)]$, then $y = f(x)$ for some $x \in f^{-1}(B)$. But then $f(x) \in B$; that is, $y \in B$. Note that $f[f^{-1}(B)]$ can be a proper subset of B; see Fig 4.1.4.
(c) If $y = f(x)$ for some $x \in A$, then x also belongs to C, so $y \in f(C)$.
(d) We have $y = f(x)$ for some $x \in \bigcup_i A_i$ iff for at least one i, $y = f(x)$ for some $x \in A_i$, iff $y \in \bigcup_i f(A_i)$. ■

Problems for Section 4.1

In Problems 1–3, all numbers are real and all functions are from R to R.

1. If $x_n \to x$ and $y_n \to y$, show that
   (a) $x_n + y_n \to x + y$.
   (b) $x_n - y_n \to x - y$.
   (c) $x_n y_n \to xy$.
   (d) $x_n / y_n \to x/y$ (if $y \neq 0$).
2. Let f and g be continuous at x. Show that $f + g, f - g, fg$, and f/g are continuous at x. ($p = fg$ is the product of f and g, i.e.,

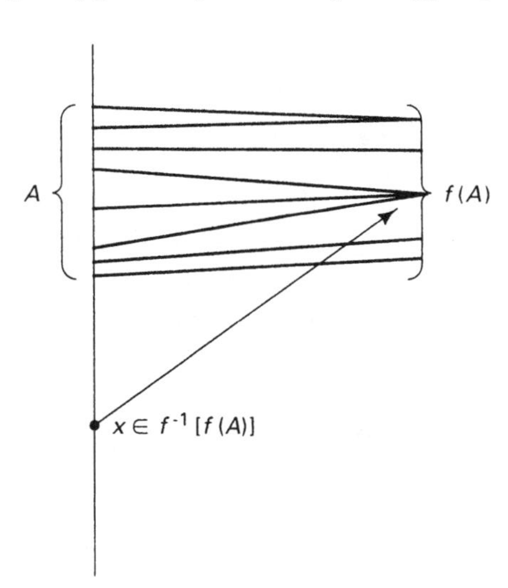

Figure 4.1.3 A Can Be a Proper Subset of $f^{-1}[f(A)]$

Figure 4.1.4 $f[f^{-1}(B)]$ Can Be a Proper Subset of B

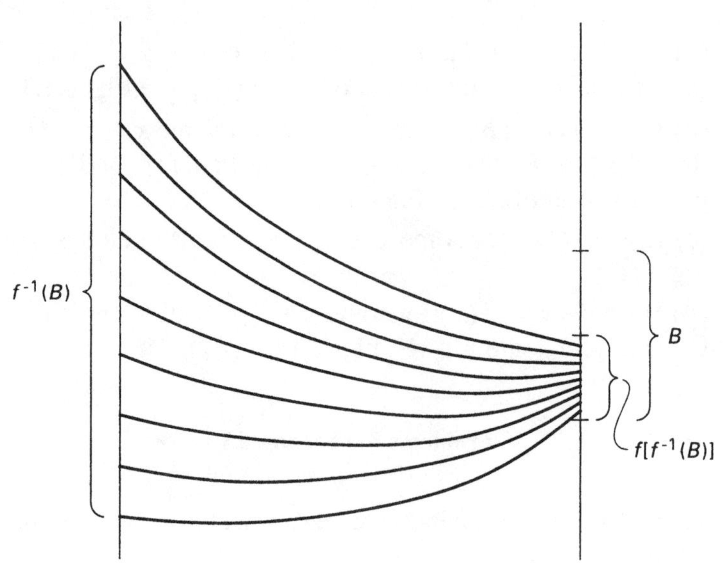

$p(y) = f(y)g(y)$, and $q = f/g$ is the quotient of f and g, i.e., $q(y) = f(y)/g(y)$. Assume $g(x) \neq 0$, so that $q(y)$ will be defined for y sufficiently close to x.)

3. Show that a polynomial ($f(x) = a_0+a_1x+\cdots+a_nx^n$) is continuous everywhere.

4. If $f: \Omega \to \Omega'$ and $\{A_i\}$ is an arbitrary family of subsets of Ω, is it true that

$$f\left(\bigcap_i A_i\right) = \bigcap_i f(A_i)?$$

5. Let A be the set of real numbers x such that $\sin x = c$, where c is a constant between -1 and $+1$. Show that A is closed. Can you identify a theorem of which this is a special case?

4.2 CONTINUITY AND COMPACTNESS

Continuous functions on compact sets have a number of special properties; in particular, they are bounded and attain a maximum and a minimum value. These results are consequences of the following fundamental theorem.

4.2.1 THEOREM. *Let $f\colon K \to \Omega'$, where K is a compact subset of the metric space Ω. If f is continuous on K, the image $f(K)$, that is, $\{f(x) : x \in K\}$, is a compact subset of Ω'. For short, the continuous image of a compact set is compact.*

Proof. Let $f(K) \subseteq \bigcup_i G_i$, where the G_i are open subsets of Ω'. Then

$$\begin{aligned} K &\subseteq f^{-1}[f(K)] && \text{by Theorem 4.1.8(a)} \\ &\subseteq f^{-1}\left(\bigcup_i G_i\right) && \text{by Theorem 4.1.5} \\ &= \bigcup_i f^{-1}(G_i) && \text{by Theorem 4.1.5.} \end{aligned}$$

By compactness, there is a finite subcovering, say

$$K \subseteq \bigcup_{j=1}^{n} f^{-1}(G_j).$$

But then

$$f(K) \subseteq \bigcup_{j=1}^{n} f[f^{-1}(G_j)] \qquad \text{by Theorem 4.1.8(c), (d).}$$

Since $f[f^{-1}(G_j)] \subseteq G_j$ by Theorem 4.1.8(b), we have a finite subcovering of the original open covering of $f(K)$. ■

4.2.2 COROLLARY. *If f is a continuous function on the compact set K, then f is bounded (in other words, $f(K)$ is bounded set). If f is real-valued, then f attains a maximum and a minimum on K.*

Proof. By Theorem 4.2.1, $f(K)$ is compact, and therefore closed and bounded by Theorem 2.2.4. In the real case, let s be the sup of $f(K)$, and let $t = \inf f(K)$. By Theorem 2.4.3, s and t belong to $f(K)$. If $f(x) = s$ and $f(y) = t$, then f attains a maximum at s and a minimum at t. ■

If the hypothesis of compactness is dropped, it is easy to produce continuous functions having no maximum or minimum value. For example, consider $f(x) = x$ on (0,1). Of course, in this case there is a natural extension of f to the compact set [0,1], and f has a maximum and a minimum on the larger domain. However, no such extension is possible for $g(x) = (1/x)\sin(1/x)$, $0 < x < 1$. The question of extension of continuous functions will be studied later in this section; it is closely connected with the idea of uniform continuity, which we now discuss.

We first look at the epsilon-delta description of continuity (see Theorem 4.1.2) for the function $f(x) = 1/x, x > 0$. Given $\epsilon > 0$ and $x > 0$, we wish to find $\delta > 0$ such that if $|x - y| < \delta$ then $|f(x) - f(y)| < \epsilon$; that is, $|x - y| < xy\epsilon$. Our first try is $\delta = \frac{1}{2}x^2\epsilon$, which is less than $xy\epsilon$ if $x < 2y$. Now if $|x - y| < \frac{1}{2}x$ (so that $y > \frac{1}{2}x$; remember that x, $y > 0$ here), then

$$|x - y| < \frac{1}{2}x^2\epsilon \implies |f(x) - f(y)| < \epsilon.$$

Thus, our final choice of δ is $\min(\frac{1}{2}x^2\epsilon, \frac{1}{2}x)$.

Our aim is not to develop techniques for finding a δ for a given ϵ, but to point out that δ may depend on x as well as ϵ. To force $1/y$ to be within distance ϵ of $1/x$, we must choose y closer to x if x is near 0 than if x is far from 0. This can be verified by looking at a picture of $1/x$; it is harder to get $1/y$ close to $1/x$ near 0 because the derivative (the slope of the tangent to the curve) is $-1/x^2$, which approaches $-\infty$ as $x \to 0$. If we can use the same δ for all x in a given set, the function is said to be uniformly continuous on that set.

4.2.3 Definition

Let $f: \Omega \to \Omega'$, where Ω and Ω' are metric spaces, and let E be a subset of Ω. We say that f is *uniformly continuous* on E if for every $\epsilon > 0$ we can find $\delta > 0$, where δ depends on ϵ *but not on* x, such that whenever x and y are points of E with $d(x, y) < \delta$, we have $d(f(x), f(y)) < \epsilon$.

For example, let $f(x) = x^2$, $0 \le x \le 1$. Then given $\epsilon > 0$, the δ for $x = 1$ works for all x in [0,1]. The key idea is that the maximum slope occurs at $x = 1$ (see Fig. 4.2.1).

Uniform continuity is a *global* concept, referring to the behavior of f on an entire set, as opposed to the *local* concept of continuity at a particular point.

In some cases, a continuous function is automatically uniformly continuous, as the following result shows.

4.2.4 THEOREM. *If f is continuous on the compact set K, then f is uniformly continuous on K.*

Proof. Fix $\epsilon > 0$. If $x \in K$, we find $\delta(x) > 0$ such that whenever $y \in K$ and $d(x,y) < \delta(x)$, we have $d(f(x), f(y)) < \epsilon/2$. The intuition behind the argument is as follows. Since K is covered by the balls $B_{\delta(x)}(x)$, compactness yields a finite subcovering. If x and y are close enough, they are likely to be in the same ball of the subcovering, say $B_{\delta(x_i)}(x_i)$. Then $d(x,x_i) < \delta(x_i)$, $d(x_i,y) < \delta(x_i)$, so that $d(f(x), f(x_i)) < \epsilon/2$, $d(f(x_i), f(y)) < \epsilon/2$. Consequently, $d(f(x), f(y)) < \epsilon$. But if x and y are not in the same ball, there is a technical difficulty, which we resolve by using balls of radius $\delta(x)/2$.

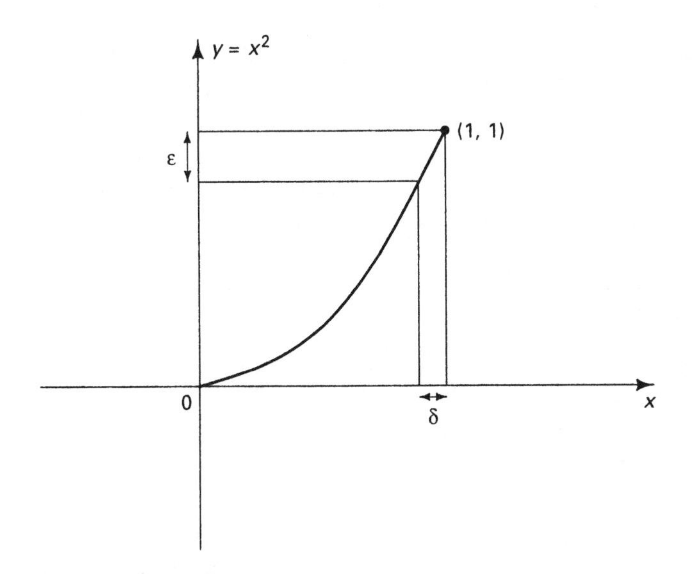

Figure 4.2.1 Uniform Continuity

To formalize, note that since $x \in B_{\delta(x)/2}(x)$, we have

$$K \subseteq \bigcup_{x \in K} B_{\delta(x)/2}(x).$$

By compactness, there is a finite subcovering, say

$$K \subseteq \bigcup_{i=1}^{n} B_{\delta(x_i)/2}(x_i).$$

Let $\delta = \frac{1}{2} \min_{1 \le i \le n} \delta(x_i) > 0$. If $x, y \in K$ and $d(x, y) < \delta$, then x belongs to some $B_{\delta(x_i)/2}(x_i)$, so $d(x, x_i) < \frac{1}{2}\delta(x_i) < \delta(x_i)$, and hence $d(f(x), f(x_i)) < \epsilon/2$. Furthermore,

$$d(x_i, y) \le d(x_i, x) + d(x, y) < \frac{1}{2}\delta(x_i) + \delta \le \frac{1}{2}\delta(x_i) + \frac{1}{2}\delta(x_i) = \delta(x_i),$$

and therefore $d(f(x_i), f(y)) < \epsilon/2$. Thus,

$$\begin{aligned} d(f(x), f(y)) &\le d(f(x), f(x_i)) + d(f(x_i), f(y)) \\ &< \frac{\epsilon}{2} + \frac{\epsilon}{2} = \epsilon. \quad \blacksquare \end{aligned}$$

(The key idea is that if $d(x, y) < \delta$, then x and y are forced to lie in the same ball $B_{\delta(x_i)}(x_i)$, so the previous argument works; see Fig. 4.2.2.)

Now consider again the function $f(x) = 1/x$, $x > 0$. Let $D = (0, \infty)$, $\Omega = [0, \infty)$, $\Omega' = R$. Then $f\colon D \to \Omega'$, and D is *dense* in Ω; that is, $\bar{D} = \Omega$. (By Theorem 1.5.1 this means that every point of Ω can be

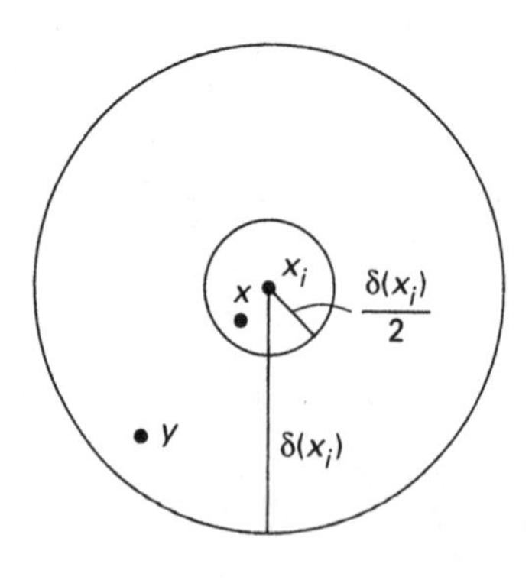

Figure 4.2.2 Proof of Theorem 4.2.4

expressed as a limit of a sequence of points in D. The most familiar result of this type is that the rationals are dense in the reals.) There is no way to extend f to a continuous mapping of Ω into Ω', for if g were such an extension then $g(0)$ would have to be $+\infty$, which is not in Ω'. (To see this, let $x \to 0$; then $g(x)$ must approach $g(0)$. But $g(x) = f(x) = 1/x$ for $x > 0$.) In fact, this shows that f cannot be uniformly continuous, as a consequence of the following result.

4.2.5 THEOREM. *Let f be a uniformly continuous mapping from D to Ω', where D is a dense subset of Ω and Ω' is a complete metric space. Then f has a unique extension to a continuous function g: $\Omega \to \Omega'$. Furthermore, g is uniformly continuous on Ω.*

Proof. If $x \in \Omega$, there is a sequence of points $x_n \in D$ with $x_n \to x$. Since $\{x_n\}$ converges, it is a Cauchy sequence (see Section 2.4.4); we claim that $\{f(x_n)\}$ is a Cauchy sequence in Ω'. For if $\epsilon > 0$, the uniform continuity of f provides $\delta > 0$ such that for all $x', y' \in D$, $d(x',y') < \delta \implies d(f(x'), f(y')) < \epsilon$. Since $d(x_n, x_m) < \delta$ for n and m sufficiently large, we have $d(f(x_n), f(x_m)) < \epsilon$ for large n and m, as desired. By completeness of Ω', $f(x_n)$ approaches a limit L; we define $g(x) = L$. We must verify that L does not depend on the particular sequence $\{x_n\}$ so that g is well defined.

Suppose $x_n \to x, y_n \to x$ with $f(x_n) \to L$ and $f(y_n) \to M$. Then

$$d(L, M) \leq d(L, f(x_n)) + d(f(x_n), f(y_n)) + d(f(y_n), M).$$

The first and third terms on the right approach 0 by hypothesis, and the second term approaches 0 because $x_1, y_1, x_2, y_2, x_3, y_3, \ldots$ is a Cauchy sequence. Thus, $d(L, M) = 0$, so $L = M$ and g is well defined.

Note that g is actually an extension of f. If $x \in D$, take $x_n = x$ for all n; then $f(x_n) \equiv f(x)$, so $g(x) = f(x)$.

Given $\epsilon > 0$, let $\delta > 0$ be such that if $x, y \in D$ with $d(x,y) < \delta$, we have $d(f(x), f(y)) < \epsilon/3$. Now consider points x, y in Ω with $d(x,y) < \delta$. Suppose we have sequences $\{x_n\}$, $\{y_n\}$ in D with $x_n \to x$, $y_n \to y$ so that $f(x_n) \to g(x)$, $f(y_n) \to g(y)$. Then

$$\begin{aligned} d(g(x), g(y)) \leq d(g(x), f(x_n)) &+ d(f(x_n), f(y_n)) \\ &+ d(f(y_n), g(y)). \end{aligned}$$

If $d(x, y) < \delta$, then $d(x_n, y_n) < \delta$ eventually, so $d(f(x_n), f(y_n)) < \epsilon/3$ for all sufficiently large n. It follows that $d(g(x), g(y)) < 3(\epsilon/3) = \epsilon$, proving that g is uniformly continuous on Ω.

To prove uniqueness, suppose h_1 and h_2 are continuous on Ω, and $h_1 = h_2$ on D. If $x \in \Omega$, let $x_n \in D$, $x_n \to x$. Since $h_1(x_n) = h_2(x_n)$ for all n, we have $h_1(x) = h_2(x)$ by continuity. Thus, $h_1 = h_2$ on Ω. ■

Problems for Section 4.2

1. Let $f(x) = x \ln x$, $x > 0$. Can f be defined at 0 so as to be continuous on the extended domain $[0, \infty)$?
2. Let $f: E \to R$, where E is a bounded subset of R^p. If f is uniformly continuous on E, use Theorem 4.2.5 to show that the image $f(E)$ is bounded.
3. Do Problem 2 without invoking Theorem 4.2.5. (Use the definition of uniform continuity to show that if $\epsilon > 0$, then E, hence $f(E)$, can be covered by a finite number of open balls of diameter less than ϵ.)
4. Let $\{x_n\}$ be a sequence in the compact subset K of the metric space Ω. If $x \in \Omega$ and x_n does *not* converge to x, show that $\{x_n\}$ has a subsequence converging to a limit $y \in K$ with $y \neq x$.
5. Apply Problem 4 to prove the following result. Let $f: K \to \Omega'$, where K is a compact susbet of the metric space Ω. If f is continuous on K and one-to-one, i.e., $f(x) = f(y)$ implies $x = y$, show that f has a continuous inverse on $f(K)$. In other words, if $x_1, x_2, \ldots \in K, x \in K$, and $f(x_n) \to f(x)$, then $x_n \to x$.
6. Give an example of a continuous function f and a closed set A such that $f(A)$ is not closed.

4.3 TYPES OF DISCONTINUITIES

It is useful to study the behavior of a function at a point where it is discontinuous. The most familiar situation occurs in the real case when f "jumps" at a point x. As t approaches x from the right (notation $t \to x^+$), $f(t)$ approaches a limit denoted by $f(x^+)$, and as t approaches

x from the left (notation $t \to x^-$), $f(t)$ approaches a limit $f(x^-)$. If there is a jump, then $f(x^-) \neq f(x^+)$.

Formally, f is said to have a *simple discontinuity* or a *discontinuity of the first kind* at x if f is discontinuous at x but $f(x^+)$ and $f(x^-)$ both exist. Within discontinuities of the first kind we have the following subcategories:

> *removable* or *point* discontinuity: $f(x^-) = f(x^+)$ finite
> *jump* discontinuity: $f(x^-) \neq f(x^+)$, both finite
> *infinite* discontinuity: $f(x^-), f(x^+)$ not both finite

(see Fig. 4.3.1).

For example, if

$$f(x) = \frac{\sin x}{x}, \quad x \neq 0, \qquad f(0) = 37,$$

then f has a removable discontinuity at $x = 0$. If

$$f(x) = 0, \quad x < 0, \qquad f(x) = 1, \quad x > 0, \qquad f(0) = 175,$$

then f has a jump discontinuity at $x = 0$. If

$$f(x) = 0, \quad x < 0, \qquad f(x) = \frac{1}{x}, \quad x > 0,$$

then f has an infinite discontinuity at $x = 0$.

To avoid confusion, we must say precisely what we mean by a statement of the form $\lim_{t \to x} f(t) = A$. The basic idea is that $f(x)$ *is not considered* in the limit statement. One way of describing the statement is to say that for all sequences t_n converging to x, *with t_n never equal to x*, we have $f(t_n) \to A$. ($A = \pm\infty$ is allowed in this case.) In epsilon-delta terms (assuming A finite), if $\epsilon > 0$ there is a $\delta > 0$ such that if $t \neq x$ and $d(t,x) < \delta$, we have $d(f(t),A) < \epsilon$ (Problem 3). Similarly, in the case where f is defined on an interval of reals, $\lim_{t \to x^+} f(t) = B$ (that is, $f(x^+) = B$) means that whenever $t_n \to x$, with $t_n > x$ for

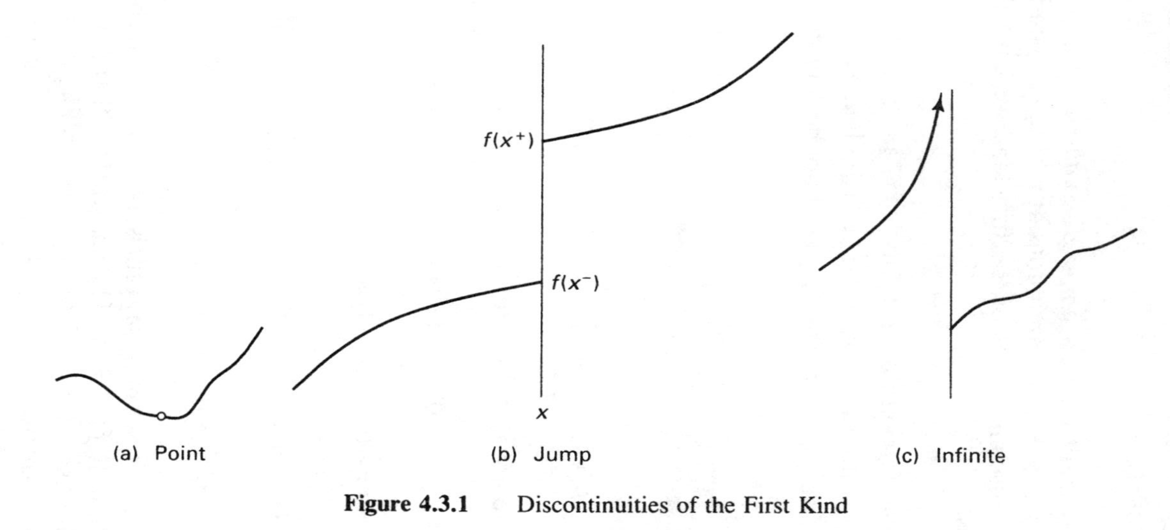

Figure 4.3.1 Discontinuities of the First Kind

all n, we have $f(t_n) \to B$. Equivalently, if $\epsilon > 0$, we can find $\delta > 0$ such that if $x < t < x + \delta$, then $d(f(t), B) < \epsilon$. In the definition of $\lim_{t \to x^-} f(t) = C$ (that is, $f(x^-) = C$), we require that $t_n < x$ for all n, or $x - \delta < t < x$ in the epsilon-delta statement. It follows that $\lim_{t \to x} f(t) = A$ if and only if $f(x^+) = f(x^-) = A$; we need not have $f(x) = A$, and in fact f might not even be defined at x (see Fig. 4.3.2). Similarly, if f has a jump discontinuity at x, $f(x)$ need not coincide with either $f(x^+)$ or $f(x^-)$. The statement that the limits $f(x^+)$ and $f(x^-)$ exist and coincide with $f(x)$ is equivalent to continuity of f at x; a formal proof can be given by using the epsilon-delta descriptions of $f(x^+)$ and $f(x^-)$ (Problem 4).

It is not hard to give an explicit example of a function that has a *nonsimple discontinuity* (also called *discontinuity of the second kind*). Consider $f(x) = \sin(1/x)$, $x \neq 0$; $f(0) = 0$. The function f is continuous everywhere except at 0, where there is a nonsimple discontinuity. For suppose $x_n = 1/n\pi$, $n = 1, 2, \ldots$; then $x_n \to 0^+$, and $f(x_n) = 0$ for all n; hence $f(x_n) \to 0$. Now if $y_n = 2/(4n+1)\pi, n = 1, 2, \ldots$, then $y_n \to 0^+$, and $f(y_n) = 1$ for n; hence $f(y_n) \to 1$. Thus, $f(t)$ has no limit as $t \to 0^+$ (similarly, $f(t)$ has no limit as $t \to 0^-$ either), and therefore 0 is a nonsimple discontinuity of f. As x gets close to 0, $\sin(1/x)$ oscillates with ever-increasing frequency.

In fact we can produce an explicit example of a function that has nothing but discontinuities of the second kind: $g(x) = 1$ for x rational, $g(x) = 0$ for x irrational. Since any real number x can be expressed as the limit of a sequence of rationals or equally well as the limit of a sequence of irrationals, every x is a nonsimple discontinuity of g.

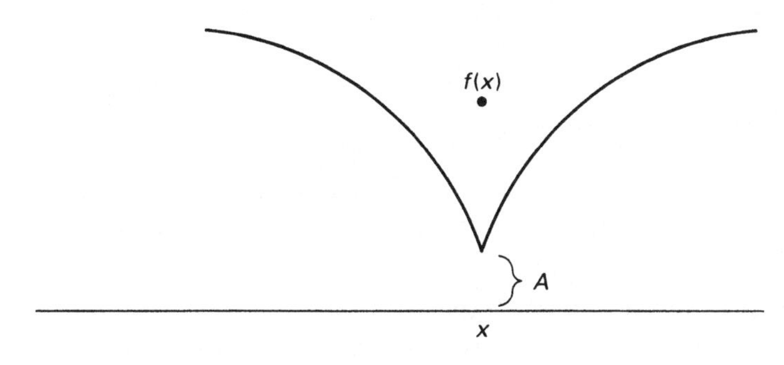

Figure 4.3.2 $\lim_{t \to x} f(t) = A$

In certain cases, discontinuities of the second kind are excluded.

4.3.1 THEOREM. *Let $f: R \to R$, and assume f monotone (that is, either increasing on R or decreasing on R). Then all discontinuities of f are jumps, and f has at most countably many discontinuities.*

Proof. Assume f increasing (the decreasing case is handled similarly, or by applying the result for increasing functions to $-f$). Intuitively, if t_n decreases to a limit x, then $f(t_n)$ converges by Theorem 2.4.6, so we expect that $f(x^+)$ exists. But how do we know that $\lim_n f(t_n)$ is the same for all sequences $\{t_n\}$? Also, what about nonmonotone $\{t_n\}$? To avoid such difficulties, we take a different approach. If $x \in R$, let $L = \inf\{f(t) : t > x\}$. Given $\epsilon > 0$, by Theorem 2.4.3 there is a real number $t_0 > x$ such that $L \leq f(t_0) < L + \epsilon$. If $t_n > x$, $t_n \to x$, then $x < t_n \leq t_0$ for all sufficiently large n. Since f is increasing, $L \leq f(t_n) \leq f(t_0) < L + \epsilon$ for large n, and therefore $f(t_n) \to L$. Thus, the limit $f(x^+)$ exists. Similarly, let $M = \sup\{f(t) : t < x\}$ and choose $t_0' < x$ such that $M - \epsilon < f(t_0') \leq M$. If $t_n < x$, $t_n \to x$, then $t_0' \leq t_n < x$ for large n, so $M - \epsilon < f(t_0') \leq f(t_n) \leq M$. Therefore, $f(t_n) \to M$, and the limit $f(x^-)$ exists. Thus, all discontinuities are simple, and since f is real-valued L and M (i.e., $f(x^+)$ and $f(x^-)$) are finite. A removable discontinuity cannot occur because f is monotone, and it follows that all discontinuities are jumps.

To show that there are at most countably many discontinuities, note that each jump determines a nonempty open interval $(f(x^-), f(x^+))$, and, by monotonicity, the open intervals are disjoint. In each interval we may choose a rational number (see Fig. 4.3.3). Since there are only countably many rationals available, there can only be countably many jumps. ■

Theorem 4.3.1 applies equally well to a function defined on an interval $I \subseteq R$; the proof is the same.

Perhaps unexpectedly, a monotone function of more than one variable can have uncountably many discontinuities (see Problem 1).

In calculus, you became familiar with the idea that a continuous function is "smooth"; in other words, its graph can be drawn without lifting

Figure 4.3.3 Proof of Theorem 4.3.1

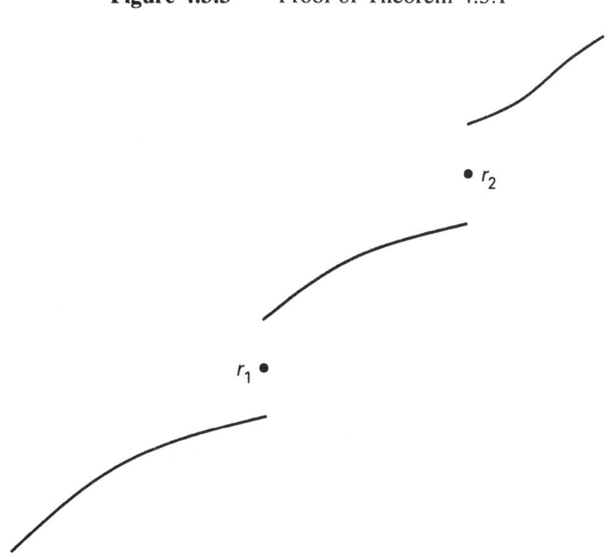

pencil from paper. As a consequence, if, say, $f(a) = 3$ and $f(b) = 10$, then as x ranges from a to b, f will take on all values between 3 and 10. We now prove a theorem to this effect.

4.3.2 Intermediate Value Theorem

Let f be a continuous, real-valued function on $[a, b]$. If $f(a) < c < f(b)$, there is an $x \in (a, b)$ such that $f(x) = c$.

Proof. The argument is rather tricky; we look at the last time that f takes a value less than c. Formally, let $A = \{t \in [a, b] : f(t) < c\}$ and let $x = \sup A$. (Note that $a \in A$, so A is not empty and therefore has a sup. Also, $A \subseteq [a, b]$, so $x \in [a, b]$; see Fig. 4.3.4.) We show that $f(x) = c$ by eliminating the other possibilities $f(x) < c$ and $f(x) > c$. By Theorem 2.4.3 there is a sequence of points $t_n \in A$ with $t_n \to \sup A = x$. Then $f(t_n) < c$ for all n, so, by continuity, $f(x) \leq c$. Thus, it remains only to dispose of the case $f(x) < c$.

If $f(x) < c$, then $x < b$, and, by continuity, if y is taken greater than x and sufficiently close to x, we have $f(y) < c$. Thus, $y \in A$, contradicting the fact that x is an upper bound of A. ■

Figure 4.3.4 Proof of Theorem 4.3.2

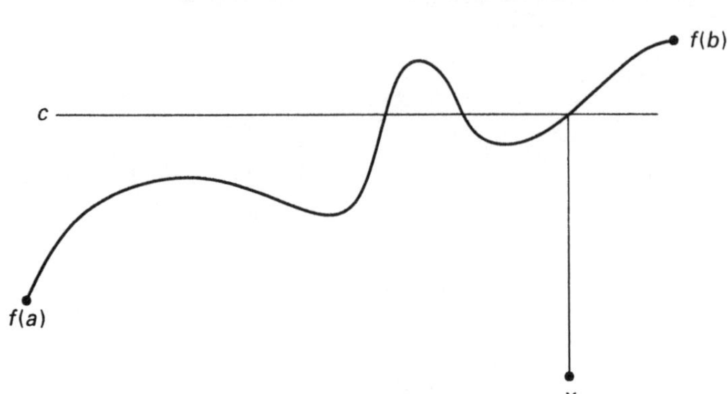

Problems for Section 4.3

1. Let $f(x,y) = 1$ if $x,y \geq 0$; $f(x,y) = 0$ elsewhere (x,y real). Show that f is increasing in the sense that if $x_1 \leq x_2$ and $y_1 \leq y_2$, then $f(x_1,y_1) \leq f(x_2,y_2)$, but f has uncountably many discontinuities.
2. Classify the discontinuities of the function given by

$$f(x) = x \sin \frac{1}{x}, \qquad x \neq 0;$$
$$f(0) = 1.$$

3. Show that the "sequence" and "epsilon-delta" characterizations of $\lim_{t \to x} f(t) = A$ are equivalent (assume A finite).
4. If $f: R \to R$, show that f is continuous at x iff $f(x^+)$ and $f(x^-)$ exist and coincide with $f(x)$.

4.4 THE CANTOR SET

This is a convenient place to introduce the Cantor set, an interesting source of examples and counterexamples. We take Ω to be the closed unit interval [0,1], and remove the "middle third" V_1; let E_1 be the set that remains:

$$V_1 = (\tfrac{1}{3}, \tfrac{2}{3}), \qquad E_1 = V_1^c = [0, \tfrac{1}{3}] \cup [\tfrac{2}{3}, 1].$$

Thus, $x \in E_1$ if and only if x can be expressed in ternary form with first digit 0 or 2. (Note that in ternary, $\frac{1}{3} = .1000\cdots = .0222\cdots \in E_1$. Although $\frac{1}{3}$ has two ternary representations, it *can* be expressed using a first digit 0, so there is no ambiguity.) We now remove the middle thirds of the intervals of E_1, and let E_2 be the set that remains:

$$V_2 = (\tfrac{1}{9}, \tfrac{2}{9}) \cup (\tfrac{7}{9}, \tfrac{8}{9}),$$
$$E_2 = E_1 \cap V_2^c = [0, \tfrac{1}{9}] \cup [\tfrac{2}{9}, \tfrac{3}{9}] \cup [\tfrac{6}{9}, \tfrac{7}{9}] \cup [\tfrac{8}{9}, 1].$$

Thus, $x \in E_2$ if and only if x can be expressed in ternary form with the first two digits 0 or 2. (For example, $\frac{6}{9} \leq x \leq \frac{7}{9}$ iff $x = .20\ldots$.)

We continue this process inductively. To make sure everyone understands what is going on, we write out one more step:

$$V_3 = (\tfrac{1}{27}, \tfrac{2}{27}) \cup (\tfrac{7}{27}, \tfrac{8}{27}) \cup (\tfrac{19}{27}, \tfrac{20}{27}) \cup (\tfrac{25}{27}, \tfrac{26}{27}),$$
$$E_3 = E_2 \cap V_3^c = [0, \tfrac{1}{27}] \cup [\tfrac{2}{27}, \tfrac{3}{27}] \cup [\tfrac{6}{27}, \tfrac{7}{27}] \cup [\tfrac{8}{27}, \tfrac{9}{27}]$$
$$\cup [\tfrac{18}{27}, \tfrac{19}{27}] \cup [\tfrac{20}{27}, \tfrac{21}{27}] \cup [\tfrac{24}{27}, \tfrac{25}{27}] \cup [\tfrac{26}{27}, 1].$$

At step n we obtain a set E_n, the union of 2^n intervals each of which is of length 3^{-n}, such that $x \in E_n$ if and only if x can be expressed in ternary form using 0's and 2's in the first n digits. Since $E_n = E_{n-1} \bigcap V_n^c$, the sets E_n form a decreasing sequence. The *Cantor set* C is defined by

$$C = \bigcap_{n=1}^{\infty} E_n.$$

Thus, $x \in C$ if and only if x can be expressed in ternary form using only digits 0 and 2. Since each E_n is closed, C is a closed set.

The following properties of C may be established.

4.4.1 THEOREM

(a) C is uncountable.

(b) The length of C is 0. (We have not defined the length of an arbitrary set of reals; this is done in measure theory. However, if we are given $\epsilon > 0$, C can be shown to be a subset of a finite union of intervals whose total length is less than ϵ. Thus, if length is defined in any reasonable way, C must have length 0.)

(c) Every point of C is a limit point of C. (A set with this property is sometimes called "perfect," although this terminology seems rather extravagant.)

(d) C is totally disconnected; that is, if $x_0, x_1 \in C$ with $x_0 < x_1$, there is a point $x_2 \in (x_0, x_1)$ with $x_2 \notin C$.

(e) C is nowhere dense; that is, the closure of C (which coincides with C since C is closed) has no interior; in other words, there are no open subsets of C except the empty set.

Proof

(a) The points of C are in one-to-one correspondence with the points in [0,1] that can be represented in binary form using the digits 0 and 1—in other words, with all points of [0,1].

(b) Note that $C \subseteq E_n$, and the length of E_n is $(\frac{2}{3})^n \to 0$.

(c) If $x = .a_1a_2a_3 \cdots \in C$ (in ternary form using digits 0 and 2), let $x_n = .a_1a_2 \ldots a_n a'_{n+1} 0\, 0 \ldots$, where we take $a'_{n+1} = 0$ if $a_{n+1} = 2$, and $a'_{n+1} = 2$ if $a_{n+1} = 0$. Then $x_n \in C, x_n \neq x$, and $x_n \to x$. Thus, x is a limit point of C.

(d) Examine the ternary expansions of x_0 and x_1. Since $x_0 < x_1$, for some n we have

$$x_0 = .a_1 \ldots a_n 0 \ldots, \qquad x_1 = .a_1 \ldots a_n 2 \ldots .$$

Let $x_2 = .a_1 \ldots a_n 11 \ldots$. Then $x_2 \notin C$ and $x_0 < x_2 < x_1$.

(e) If $(a,b) \subseteq C$, let $x_0, x_1 \in (a,b)$ with $x_0 < x_1$. By (d) we can find $x_2 \in (x_0, x_1)$ with $x_2 \notin C$, contradicting $x_2 \in (a,b) \subseteq C$. ∎

4.4.2 Remarks

The Cantor set can be used to construct continuous functions with rather unusual properties. Here is one example. If A is a nonempty subset of the metric space Ω and $x \in \Omega$, minimizing the distance $d(x,y)$ as y ranges over A would seem to produce a reasonable notion of *distance from a point to a set*. Formally, we define

$$d(x,A) = \inf\{d(x,y) : y \in A\}.$$

Now

$$d(x,z) \leq d(x,y) + d(y,z) \qquad \text{for all} \quad x,y,z \in \Omega;$$

thus

$$d(x,A) \leq d(x,y) + d(y,z) \qquad \text{for all} \quad x,y \in \Omega, \ z \in A.$$

Take the inf over $z \in A$ to obtain $d(x,A) \leq d(x,y) + d(y,A)$; by symmetry,

$$|d(x,A) - d(y,A)| \leq d(x,y).$$

Thus, the function defined by $f(x) = d(x,A)$, $x \in \Omega$, is uniformly continuous. If A is closed and $x \notin A$, then (since A^c is open), $B_r(x) \subseteq A^c$ for some $r > 0$. If $y \in A$, then $d(x,y) \geq r$; hence, $d(x,A) \geq r > 0$. (Of course, if $x \in A$, then $d(x,A) = 0$ whether or not A is closed).

Now let A be the Cantor set C and take Ω to be $[0,1]$. Since C is closed, $f(x) = 0$ if and only if $x \in C$. By Theorem 4.4.1(a), f has uncountably many zeros, and by Theorem 4.4.1(d), f is never identically 0 on any open subinterval of $[0,1]$.

Problems for Section 4.4

1. Let Ω be a metric space with distance function d. Let $f(x) = d(x,y)$, $x \in \Omega$, where y is any fixed element of Ω. Show that f is uniformly continuous on Ω.

2. Indicate how to modify the construction of the Cantor set so that a set of positive length is obtained.
3. If A is compact (and nonempty) and x is arbitrary, show that there is a point $y_0 \in A$ such that

$$d(x, y_0) = d(x, A).$$

4. In Problem 3, show that "compact" can be replaced by "closed" if $\Omega = R^p$.

REVIEW PROBLEMS FOR CHAPTER 4

1. Give an example of a function $f\colon R \to R$ that has a nonsimple discontinuity at $x = 2$ but no other discontinuities.
2. Give an example of a set A and a point $x \notin A$ such that $d(x, A) = 0$.
3. Let $f(x) = x$ if x is rational; $f(x) = 0$ if x is irrational. Is f discontinuous everywhere?
4. Let $f\colon A \to R$, where A is the set of rational numbers x such that $0 < x < 7$. If f is uniformly continuous on A, show that f is bounded.
5. If $f(x) = x^3 + 2x + 2$, show that $f(x) = 0$ for some x in $(-1, 0)$.
6. Let $f(x) = \ln x$; is f uniformly continuous on (0,1]?
7. Let A be the set of all real numbers x such that e^{x^2} is not an integer. Show that A is open.
8. Give an example of a countably infinite nowhere dense subset of R.
9. Let $f(x) = 1$ if $x \geq 0$; $f(x) = 0$ if $x < 0$. Find a closed set $C \subseteq R$ such that $f^{-1}(C)$ is not closed.
10. Classify the discontinuities of the function given by

$$f(x) = \frac{e^{1/x}}{(x-1)^2} \qquad \text{if } x \neq 0,\ x \neq 1;$$

$$f(0) = f(1) = 0.$$

5

DIFFERENTIATION

5.1 THE DERIVATIVE AND ITS BASIC PROPERTIES

In this section we examine the differentiation process and obtain the Mean Value Theorem, which has several basic applications, including L'Hospital's rule and Taylor's formula with remainder.

5.1.1 Definition and Comments

If $f\colon (a,b) \to R$ and $x \in (a,b)$, the *derivative of f at x* is defined by

$$f'(x) = \lim_{h \to 0} \frac{f(x+h) - f(x)}{h},$$

provided the limit exists and is finite. In this case we say that f is *differentiable* at x.

We will not derive the familiar calculus formulas for the derivative of a sum, product, or quotient of two functions, but the *chain rule* is worth looking at. Here we assume that f is differentiable at x (when making a statement of this type, we always assume that f is defined at least on some open interval containing x) and g is differentiable at $y = f(x)$.

If h is the composition of f and g, that is, $h(t) = g(f(t))$, we wish to show that h is differentiable at x and

$$h'(x) = g'(f(x))f'(x).$$

There is a rather intuitive argument that almost works. We use Δx instead of h for the increment in x, and write $\Delta y = f(x + \Delta x) - f(x)$, so $y + \Delta y = f(x + \Delta x)$. Let $\Delta z = h(x + \Delta x) - h(x)$; we then have

$$\begin{aligned}\Delta z &= g(f(x + \Delta x)) - g(f(x))\\ &= g(y + \Delta y) - g(y).\end{aligned}$$

Thus,

$$\text{(1)} \quad \frac{\Delta z}{\Delta x} = \frac{\Delta z}{\Delta y}\frac{\Delta y}{\Delta x} = \frac{[g(y + \Delta y) - g(y)]}{\Delta y}\frac{[f(x + \Delta x) - f(x)]}{\Delta x}.$$

We would simply like to let $\Delta x \to 0$ to obtain the desired result. However, there is a problem of division by 0. Although Δx is restricted to be nonzero by definition of $f'(x)$ (see the remarks at the beginning of Section 4.3), we have no guarantee that Δy will be nonzero; since Δy is determined by Δx, we have no control over it. One possible solution is to *define* $[g(y + \Delta y) - g(y)]/\Delta y$ to be $g'(y)$ if $\Delta y = 0$. In this case, $f(x + \Delta x) - f(x) = \Delta y = 0$, so the right side of (1) is 0; also $\Delta z = g(y + \Delta y) - g(y) = 0$, so the left side is 0 as well. Now if we make Δx sufficiently small, $[f(x + \Delta x) - f(x)]/\Delta x$ will be arbitrarily close to $f'(x)$. Furthermore, if we knew that $\Delta x \to 0$ implies $\Delta y \to 0$, we would be able to conclude that $[g(y + \Delta y) - g(y)]/\Delta y$ approaches $g'(f(x))$, proving the chain rule. But the statement that $\Delta x \to 0$ implies $\Delta y \to 0$ is just the assertion that differentiability implies continuity, which we now prove.

5.1.2 THEOREM. *If f is differentiable at x, then f is continuous at x.*

Proof. Let $\Delta y = f(x + \Delta x) - f(x)$; then

$$\begin{aligned}\Delta y = \frac{\Delta y}{\Delta x}\,\Delta x &= f'(x)\,\Delta x + \left[\frac{\Delta y}{\Delta x} - f'(x)\right]\,\Delta x\\ &\to 0 \quad \text{as} \quad \Delta x \to 0:\end{aligned}$$

Thus, $f(x + \Delta x) \to f(x)$ as $\Delta x \to 0$. ■

The familiar results on local maxima and minima may now be established.

5.1.3 THEOREM. *Let f be differentiable at x, and assume that f has either a local maximum at x (that is, $f(x+h) \le f(x)$ for all sufficiently small h) or a local minimum at x ($f(x+h) \ge f(x)$ for all sufficiently small h). Then $f'(x) = 0$.*

Proof. Suppose f has a local maximum at x; then

$$\frac{f(x+h) - f(x)}{h} \begin{cases} \le 0 & \text{for} \quad h > 0, \\ \ge 0 & \text{for} \quad h < 0. \end{cases}$$

Let $h \to 0$ to conclude that $f'(x)$ is both ≤ 0 and ≥ 0; hence $f'(x) = 0$. The argument for a local minimum is similar. ■

We now prove a result that is perhaps the most useful of all differentiation theorems.

5.1.4 Mean Value Theorem

Let f: $[a, b] \to R$, and assume that f is continuous on $[a, b]$ and differentiable on (a, b). Then for some $x \in (a, b)$ we have

$$\frac{f(b) - f(a)}{b - a} = f'(x).$$

Thus, at x, the slope of the tangent to the curve is the same as the slope of the chord joining $(a, f(a))$ to $(b, f(b))$ (see Fig. 5.1.1). In the special case $f(a) = f(b) = 0$, we have $f'(x) = 0$ for some $x \in (a, b)$, so that the curve has a horizontal tangent. This is known as *Rolle's Theorem* (see Fig. 5.1.2). (Intuitively, if the average speed of a car is 60 mph, there is an instant at which the speedometer reads exactly 60. If the car is slower at the beginning, it must be faster later.)

Proof. We consider Rolle's Theorem first. If f is identically 0, there is nothing to prove, so assume $f(t) > 0$ for some $t \in (a, b)$. Since f is continuous on the compact set $[a, b]$, by Corollary 4.2.2, f attains a maximum at some x, necessarily in (a, b). (If $f(t) < 0$ for some t, we obtain a minimum at some point of (a, b).) By Theorem 5.1.3, $f'(x) = 0$.

Figure 5.1.1 Mean Value Theorem

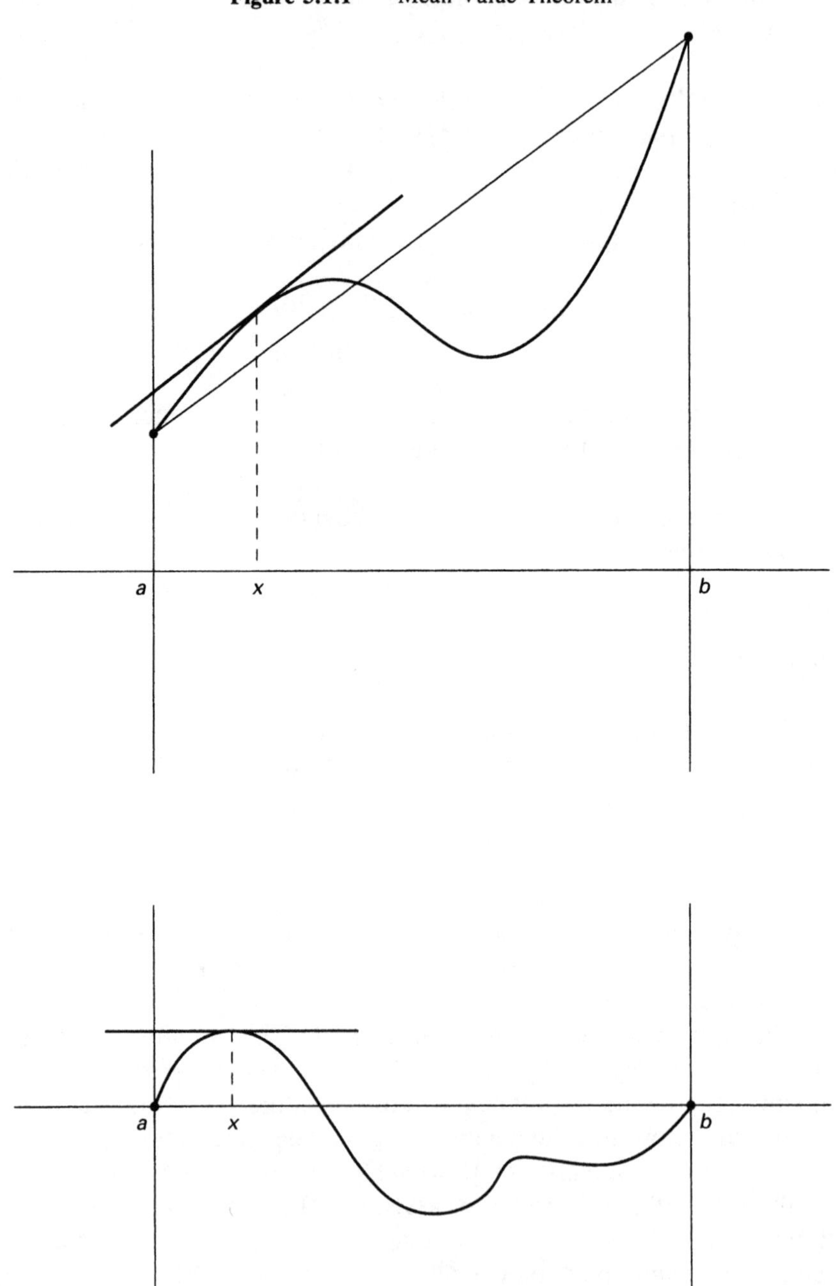

Figure 5.1.2 Rolle's Theorem

The general case may be reduced to Rolle's Theorem as follows. Define the function g by subtracting a linear function from f as follows:

$$g(t) = f(t) - f(a) - \frac{f(b) - f(a)}{b - a}(t - a).$$

Then g is continuous on $[a, b]$, differentiable on (a, b), and $g(a) = g(b) = 0$. Furthermore,

$$g'(t) = f'(t) - \frac{f(b) - f(a)}{b - a}.$$

By Rolle's Theorem, $g'(x) = 0$ for some $x \in (a, b)$, and the result follows. ■

5.1.5 COROLLARY. *Let f be differentiable on the open interval I.*

(a) If $f' \geq 0$ on I, then f is increasing on I; that is, if $a, b \in I$, $a < b$, then $f(a) \leq f(b)$. If $f' > 0$ on I, then f is strictly increasing on I.

(b) If $f' \leq 0$ on I, then f is decreasing on I; that is, if $a, b \in I$; $a < b$, then $f(a) \geq f(b)$. If $f' < 0$ on I then f is strictly decreasing on I.

(c) If $f' = 0$ on I, then f is constant on I.

Proof. If $a, b \in I$, $a < b$, then, by Theorem 5.1.4, $f(b) - f(a) = (b - a)f'(x)$ for some $x \in (a, b)$; the result follows. ■

The following generalization of Theorem 5.1.4 will be needed.

5.1.6 Generalized Mean Value Theorem

Assume that f and g are continuous real-valued functions on $[a, b]$, both differentiable on (a, b). We assume that g' is never 0 on (a, b); hence, $g(b) - g(a) \neq 0$ by Theorem 5.1.4. Then for some $x \in (a, b)$ we have

$$\frac{f(b) - f(a)}{g(b) - g(a)} = \frac{f'(x)}{g'(x)}.$$

When $g(t) = t$, this reduces to Theorem 5.1.4.

[Intuitively, if car f covers four times as much distance as car g, there is an instant when its speedometer reads four times as much. (If $f'/g' < 4$ at the beginning, it must be > 4 later, to compensate.)]

Proof. Consider the function h defined by

$$h(t) = f(t) - f(a) - \frac{f(b) - f(a)}{g(b) - g(a)}[g(t) - g(a)].$$

Then h is continuous on $[a, b]$, differentiable on (a, b), and $h(a) = h(b) = 0$. By Rolle's Theorem, $h'(x) = 0$ for some $x \in (a, b)$, and the theorem is proved. ■

Problems for Section 5.1

1. Let $f(x) = x^2 \sin(1/x), x \neq 0$; $f(0) = 0$. Show that f is differentiable everywhere.
2. If $f\colon R \to R$ and for some $r > 0$ and all $x, y \in R$ we have

   $$|f(x) - f(y)| \leq |x - y|^{1+r},$$

   show that f is constant.
3. Let $p(x) = a_0 + a_1 + \cdots + a_n x^n$ be a polynomial with real coefficients. If all roots of p are real, show that all roots of the derivative p' are real also.
4. If the hypothesis that g' is never 0 on (a, b) is dropped from the statement of the Generalized Mean Value Theorem, the result is no longer true. Can you give an explicit counterexample (preferably one with $g(b) - g(a) \neq 0$ and $f'(x_0) = 0$ whenever $g'(x_0) = 0$, with $f'(x)/g'(x) \to$ finite limit as $x \to x_0$)?

5.2 ADDITIONAL PROPERTIES OF THE DERIVATIVE; SOME APPLICATIONS OF THE MEAN VALUE THEOREM

Let's examine the assumption we made in Theorem 5.1.6 that g' is never 0 on (a, b). It is natural to expect that either $g' > 0$ on (a, b) or $g' < 0$ on (a, b), for if $g'(c) < 0 < g'(d)$ then g' would be 0 somewhere between c and d, which is impossible. In presenting this argument, we are applying the Intermediate Value Theorem 4.3.2 to g', but we have a problem because g' need not be continuous. However, the result holds anyway, as we now show.

5.2.1 Intermediate Value Theorem for Derivatives

Assume f is differentiable on the open interval I and that $a, b \in I$, $a < b$, with $f'(a) < c < f'(b)$. There is a point $x \in (a,b)$ such that $f'(x) = c$.

Proof. Let $g(t) = f(t) - ct$; then $g'(a) = f'(a) - c < 0$, and $g'(b) = f'(b) - c > 0$. Since $[g(a+h) - g(a)]/h \to g'(a)$ as $h \to 0$, we have $g(a+h) - g(a) < 0$ for $h > 0$ and sufficiently small. Similarly, $[g(b+h) - g(b)]/h \to g'(b)$ as $h \to 0$; hence, $g(b+h) - g(b) < 0$ for h *negative* and sufficiently small. Thus, we can find points $t_1, t_2 \in (a,b)$ such that $g(t_1) < g(a)$ and $g(t_2) < g(b)$. The point of doing this is to conclude that if x minimizes $g(t)$, $a \le t \le b$ (the existence of x is guaranteed by Corollary 4.2.2), then we must have $a < x < b$, so by Theorem 5.1.3, $g'(x) = 0$; in other words, $f'(x) = c$. ■

5.2.2 COROLLARY. *If f is differentiable on an open interval containing x_0, then f' cannot have a simple discontinuity at x_0.*

Proof. Intuitively, if f' jumps at x_0, a typical situation might be $f'(x) = 0, x < x_0$; $f'(x) = k, x \ge x_0$. By considering the cumulative area under f', we conclude that $f(x) = 0, x < x_0$; $f(x) = k(x - x_0)$, $x \ge x_0$. But then f is not differentiable at x_0 (see Fig. 5.2.1).

For the formal proof, first assume f' has a simple discontinuity at x_0 with $f'(x_0^-) < c < f'(x_0^+)$; see Fig. 5.2.2. By changing c slightly, we

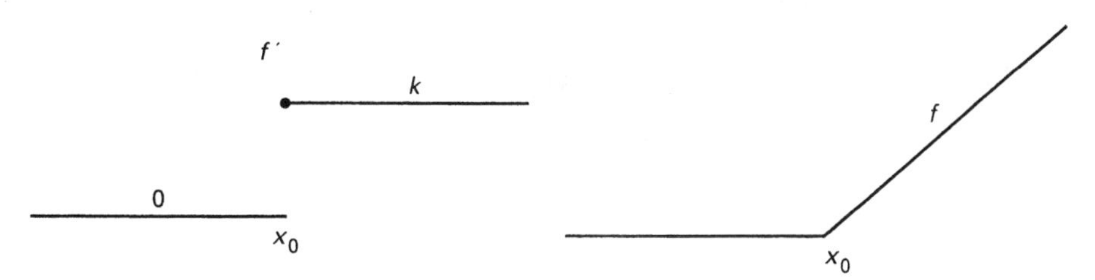

Figure 5.2.1 Intuitive Argument That the Derivative Cannot Have a Simple Discontinuity

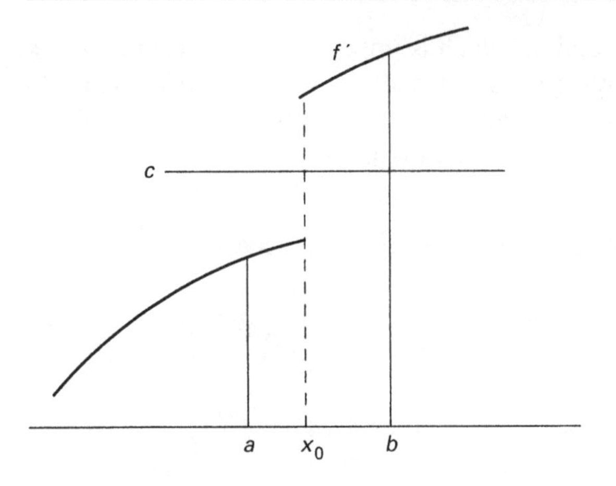

Figure 5.2.2 Proof of Corollary 5.2.2

may assume $c \neq f'(x_0)$. By definition of a simple discontinuity, $f'(t) \to f'(x_0^+)$ as $t \to x_0$ from above, and $f'(t) \to f'(x_0^-)$ as $t \to x_0$ from below. It follows that we can find points a, b and an $\epsilon > 0$ such that $a < x_0 < b$ and

$$f'(t) \leq c - \epsilon, \qquad a \leq t < x_0,$$

$$f'(t) \geq c + \epsilon, \qquad x_0 < t \leq b.$$

Now apply Theorem 5.2.1 to obtain $x \in (a, b)$ such that $f'(x) = c$; in view of the preceding inequalities we must have $x = x_0$. But this is a contradiction because $f'(x_0) \neq c$.

There is one remaining possibility, namely, $f'(x_0^-) = f'(x_0^+) \neq f'(x_0)$, and this is covered by a similar argument (Problem 2). Thus, f' cannot have a simple discontinuity at x_0. ■

For an example of a function whose derivative has a nonsimple discontinuity, see Problem 3.

We now discuss a basic application of the Generalized Mean Value Theorem.

5.2.3 L'Hospital's Rule

In calculus, you encountered the problem of finding the limit of a quotient $f(x)/g(x)$ as $x \to a$, where $f(x)$ and $g(x)$ both approach 0 as $x \to a$ (the "0/0 case"), or $f(x)$ and $g(x)$ both approach ∞ as $x \to a$ (the "∞/∞ case"). In either case, if f and g are differentiable, with g' never 0, on (a, b), and $f'(x)/g'(x) \to L$ as $x \to a$, then $f(x)/g(x) \to L$ as $x \to a$. (In this case, L is allowed to be infinite.)

To prove the assertion, consider the 0/0 case first. Since f is differentiable, hence continuous, on (a, b), and $f(x) \to 0$ as $x \to a$, we may extend f to a continuous function on $[a, b)$, and similarly for g. If $a < x < b$, Theorem 5.1.6 yields

$$\frac{f(x)}{g(x)} = \frac{f(x) - f(a)}{g(x) - g(a)} = \frac{f'(y)}{g'(y)} \qquad \text{for some} \quad y \in (a, x);$$

let $x \to a$ to obtain $f(x)/g(x) \to L$, as desired. (Intuitively, if the speed of car f is four times the speed of car g at time a, and they start from the same position at that time, then in a small time interval beginning with a, car f will cover four times as much distance.)

In the ∞/∞ case, let $a < x < t_0 < b$, and apply Theorem 5.1.6 to obtain

$$\text{(1)} \qquad \frac{f(t_0) - f(x)}{g(t_0) - g(x)} = \frac{f'(y)}{g'(y)} \qquad \text{for some} \quad y \in (x, t_0).$$

Now observe that

$$\text{(2)} \qquad \frac{f(t_0) - f(x)}{g(t_0) - g(x)} = \frac{f(x)}{g(x)} \left[\frac{1 - f(t_0)/f(x)}{1 - g(t_0)/g(x)} \right].$$

The expression in brackets approaches 1 as $x \to a$, and it follows from (1) and the assumption that $f'(y)/g'(y) \to L$ as $y \to a$ that $f(x)/g(x) \to L$, as desired. (Formally, we may first choose t_0 so that $f'(t)/g'(t)$ is close to L for all $t \in (a, t_0)$, and then let $x \to a$.)

Our final application of the Mean Value Theorem is to the problem of expanding a function f in a power series. If we are interested in the behavior of f near $x = a$, we may attempt to represent f in the form $a_0 + a_1(x - a) + a_2(x - a)^2 + \cdots$. If such a representation is possible, the coefficient a_n must be given by $f^{(n)}(a)/n!$, where $f^{(n)}$ is the nth derivative of f (take $f^{(0)} = f$). The general problem of convergence of power series is best studied via complex variables, but we can obtain the following useful result.

5.2.4 Taylor's Formula with Remainder

Assume that $f^{(n)}$ exists on the open interval I, and let $a, b \in I$. Then

$$f(b) = f(a) + f'(a)(b-a) + \frac{f''(a)}{2!}(b-a)^2 + \cdots + \frac{f^{(n-1)}(a)}{(n-1)!}(b-a)^{n-1} + R_n,$$

where

$$R_n = \frac{f^{(n)}(x)}{n!}(b-a)^n \qquad \text{for some } x \text{ between } a \text{ and } b.$$

Proof. Since differentiability implies continuity, f and all its derivatives up to order $n-1$ are continuous on I. Assume for simplicity that $a < b$; the argument for $a > b$ is essentially the same. Let M be defined by the equation

$$f(b) = f(a) + f'(a)(b-a) + \cdots + \frac{f^{(n-1)}(a)}{(n-1)!}(b-a)^{n-1} + \frac{M(b-a)^n}{n!}.$$

We must produce $x \in (a, b)$ such that $f^{(n)}(x) = M$. Essentially, we replace a by a variable t in the preceding equation; let

$$g(t) = -f(b) + f(t) + f'(t)(b-t) + \cdots + \frac{f^{(n-1)}(t)}{(n-1)!}(b-t)^{n-1} + \frac{M(b-t)^n}{n!}, \qquad t \in I.$$

By hypothesis, g is continuous on $[a, b]$ and differentiable on (a, b). Also, $g(b) = 0$ by definition of g, and $g(a) = 0$ by definition of M.

Thus, by Theorem 5.1.4, $g'(x) = 0$ for some $x \in (a, b)$. But

$$\begin{aligned} g'(t) =& f'(t) - f'(t) + (b-t)f''(t) - (b-t)f''(t) \\ &+ \frac{(b-t)^2}{2!} f^{(3)}(t) \\ &- \frac{(b-t)^2}{2!} f^{(3)}(t) + \cdots \\ &+ \frac{f^{(n)}(t)}{(n-1)!}(b-t)^{n-1} - \frac{M(b-t)^{n-1}}{(n-1)!}. \end{aligned}$$

All terms cancel except for the last two, so if we find $x \in (a, b)$ such that $g'(x) = 0$, we must have $f^{(n)}(x) = M$, as desired. ■

Problems for Section 5.2

1. Give an example of a function g differentiable on an open interval containing $[a, b]$ such that the minimum value of $g(t)$, $a \le t \le b$, occurs at a, but $g'(a) \neq 0$. (This is the case we took pains to exclude in the proof of Theorem 5.2.1.)
2. Complete the proof of Corollary 5.2.2 by analyzing the case $f'(x_0^-) = f'(x_0^+) \neq f'(x_0)$.
3. Let $f(x) = x^2 \sin(1/x)$, $x \neq 0$; $f(0) = 0$. By Problem 1 of Section 5.1, f is differentiable everywhere. Show that f' has a nonsimple discontinuity at $x = 0$.
4. Use Taylor's formula with remainder to show that the familiar power series expansion

$$e^x = \sum_{n=0}^{\infty} \frac{x^n}{n!}$$

is valid for all x.

5. Discuss the application of the Mean Value Theorem to the problem of estimating how close b must be to a in order that $f(b)$ be within a specified degree of closeness to $f(a)$. State your assumptions clearly.

REVIEW PROBLEMS FOR CHAPTER 5

1. Let f be differentiable on R, with $f'(-1) < 0$, $f'(1) > 0$. Must f have a local maximum at some point in $(-1, 1)$? Explain.

2. Let $f(x) = e^{-1/x^2}$ if $x \neq 0$; $f(0) = 0$. Show that f is infinitely differentiable on R; that is, the nth derivative $f^{(n)}(x)$ exists for every $n = 1, 2, \ldots$ and every $x \in R$.

3. Give an example of a function f whose Taylor expansion $\sum_{n=0}^{\infty} [f^{(n)}(0)/n!]\,x^n$ converges for every $x \in R$ but does not converge to $f(x)$ except at $x = 0$.

4. Give an example of functions f: $(0,1) \to R$ and g: $(0,1) \to R$ such that f and g are differentiable everywhere, g is never 0, g' is never 0, and $f(x)/g(x) \to 1$ as $x \to 0$, but $f'(x)/g'(x) \to L \neq 1$ as $x \to 0$.

5. Give an example of a function f that is differentiable everywhere and whose derivative has a nonsimple discontinuity at $x = 1$.

6. Use Taylor's formula with remainder to show that the power series expansion

$$\sin x = x - \frac{x^3}{3!} + \frac{x^5}{5!} - \frac{x^7}{7!} + \cdots$$

is valid for all x.

7. Let f and g be differentiable on the open interval I, and let a be a point of I with $f(a) = g(a) = 0$. Consider the following "proof" of the 0/0 case of L'Hospital's Rule:
For x near a, we may write, by the Mean Value Theorem,

$$f(x) = f(a) + (x-a)f'(y) \qquad \text{and} \qquad g(x) = g(a) + (x-a)g'(z),$$

where y and z are between a and x. Now let x, hence y and z, approach a. If $f'(x)/g'(x) \to L$, it follows that $f(x)/g(x) \to L$ also.

   (a) Explain why this argument is unsound.

   (b) The proof works if an extra hypothesis is added. What is this hypothesis, and where is it used?

6

RIEMANN–STIELTJES INTEGRATION

6.1 DEFINITION OF THE INTEGRAL

In calculus you did many computational examples and applications involving the Riemann integral $\int_a^b f(x)\,dx$. In this section we examine the precise definition of a somewhat more general integral. The familiar intuitive notion of the integral as the limit of the sum of a very large number of very small quantities is central to the discussion.

6.1.1 Definitions and Comments

Let f and α be functions from $[a,b]$ to R, and let P: $a = x_0 < x_1 < \cdots < x_n = b$ be a partition of $[a,b]$. The *size* of P is defined as the largest subinterval length; that is,

$$|P| = \max\{|x_i - x_{i-1}| : 1 \le i \le n\}.$$

We assume throughout the discussion that α is increasing and f is bounded. As in the familiar process of making a rectangular approximation for the purpose of computing the area under a curve, we define

$$M_k = \sup\{f(x) : x_{k-1} \le x \le x_k\},$$
$$m_k = \inf\{f(x) : x_{k-1} \le x \le x_k\}, \qquad k = 1, 2, \ldots, n.$$

(Since f is bounded, M_k and m_k are finite.) If t_k is a point in $[x_{k-1}, x_k]$, $k = 1, 2, \ldots, n$ (the choice of the t_k may be regarded as part of the specification of P), let

$$f_k = f(t_k), \qquad k = 1, 2, \ldots, n.$$

Also define

$$\Delta\alpha_k = \alpha(x_k) - \alpha(x_{k-1}), \qquad k = 1, 2, \ldots, n.$$

The *upper sum* associated with P, f, α is defined as

$$U(P, f, \alpha) = \sum_{k=1}^{n} M_k \, \Delta\alpha_k;$$

similarly, the *lower sum* is

$$L(P, f, \alpha) = \sum_{k=1}^{n} m_k \, \Delta\alpha_k$$

and the *Riemann–Stieltjes sum* is

$$S(P, f, \alpha) = \sum_{k=1}^{n} f_k \, \Delta\alpha_k.$$

If f and α are understood, we write these sums as $U(P)$, $L(P)$, and $S(P)$. When $\alpha(x) = x$ for all x, we have $\Delta\alpha_k = x_k - x_{k-1}$; in this case, we obtain approximating sums to the Riemann integral.

If f is a bounded function such that $U(P)$ and $L(P)$ approach a common finite limit as $|P| \to 0$, we say that f is *Riemann–Stieltjes integrable* with respect to α or that the *Riemann–Stieltjes integral* of f with respect to α exists. The integral is denoted by

$$\int_a^b f(x)\, d\alpha(x) \qquad \text{or} \qquad \int_a^b f\, d\alpha.$$

If $\alpha(x) = x$, $\int_b^a f \, d\alpha$ is called the *Riemann integral* of f, denoted by $\int_a^b f(x)\,dx$.

In numerous practical examples, f is a continuous function. Before showing that the Riemann–Stieltjes integral always exists in this case, we need some preliminaries. By Theorem 4.2.4, f is uniformly continuous on $[a, b]$, so that given $\epsilon > 0$ there is a $\delta > 0$ such that whenever $|x - y| < \delta$ we have $|f(x) - f(y)| < \epsilon$. Let us prove that if α is increasing and P is a partition of size less than δ,

$$|U(P) - L(P)| \leq \epsilon[\alpha(b) - \alpha(a)]. \tag{1}$$

To establish this, observe that the assumption $|P| < \delta$ implies that $0 \leq M_k - m_k \leq \epsilon$ for all k; hence,

$$\begin{aligned} |U(P) - L(P)| &= \left| \sum_{k=1}^{n} (M_k - m_k)\, \Delta\alpha_k \right| \\ &\leq \epsilon \sum_{k=1}^{n} |\Delta\alpha_k| \\ &= \epsilon[\alpha(b) - \alpha(a)] \qquad \text{by montonicity.} \end{aligned}$$

Now consider a *refinement* of P, in other words, keep the points $x_0, \ldots, x_n$ defining P and add new points (possibly in each subinterval) to form a partition P'. If a new point y appears in the subinterval $[x_{k-1}, x_k]$ of P so that $x_{k-1} < y < x_k$, then the term $M_k[\alpha(x_k) - \alpha(x_{k-1})]$, which appears in $U(P)$, may be written as

$$M_k[\alpha(y) - \alpha(x_{k-1})] + M_k[\alpha(x_k) - \alpha(y)].$$

If $M_k' = \sup\{f(x) : x_{k-1} \leq x \leq y\}$, $M_k'' = \sup\{f(x) : y \leq x \leq x_k\}$, the corresponding expression in $U(P')$ is

$$M_k'[\alpha(y) - \alpha(x_{k-1})] + M_k''[\alpha(x_k) - \alpha(y)]$$

(note also that $M_k' \leq M_k$, $M_k'' \leq M_k$; similarly, refinement increases lower sums). The point of all this is to compare $U(P)$ with $U(P')$. Since $|M_k - M_k'| \leq \epsilon$, $|M_k - M_k''| \leq \epsilon$, we obtain, as in the proof of (1),

(2) $$|U(P) - U(P')| \leq \epsilon[\alpha(b) - \alpha(a)],$$

and, similarly,

$$|L(P) - L(P')| \leq \epsilon[\alpha(b) - \alpha(a)].$$

Finally, it follows immediately from the definitions that

(3) $$L(P) \leq S(P) \leq U(P).$$

It is convenient at this point to make a comment that we need later. Suppose P' is formed by adding only one point y to P in the subinterval $[x_{k-1}, x_k]$. The situation we are analyzing involves a *fixed* y, with the size of P (hence also the size of P') approaching 0. If f is continuous at y, then M_k, M_k', and M_k'' all approach $f(y)$ as $|P| \to 0$. Thus, $|U(P) - U(P')|$ can be made less than ϵ if $|P|$ is sufficiently small. If α is continuous at y, then $\alpha(y) - \alpha(x_{k-1})$ and $\alpha(x_k) - \alpha(y) \to 0$ as $|P| \to 0$. Since M_k, M_k', and M_k'' are all bounded by $\sup_{a \leq x \leq b} f(x)$, we can again conclude that $|U(P) - U(P')|$ can be made less than ϵ if $|P|$ is sufficiently small. (A similar analysis can be made for lower sums.) If f and α are discontinuous at the same point, $\int_a^b f\, d\alpha$ does not exist, as you should discover when doing Problem 1.

We are now ready for the main result.

6.1.2 THEOREM. *If f is continuous and α is increasing on $[a, b]$, then the Riemann–Stieltjes integral $\int_a^b f\, d\alpha$ exists. In particular, if f is continuous on $[a, b]$, then f is Riemann integrable on $[a, b]$.*

Proof. Given $\epsilon > 0$, let $\delta > 0$ be such that $|x - y| < \delta$ implies $|f(x) - f(y)| < \epsilon$. If P_1 and P_2 are arbitrary partitions with $|P_1|$ and $|P_2| < \delta$, let P be a common refinement of P_1 and P_2 (take the union of the points of P_1 and the points of P_2 to form P). By Eq. (2),

$$|U(P_1) - U(P)| \leq \epsilon[\alpha(b) - \alpha(a)]$$

and

$$|U(P_2) - U(P)| \leq \epsilon[\alpha(b) - \alpha(a)].$$

Thus,

$$|U(P_1) - U(P_2)| \leq 2\epsilon[\alpha(b) - \alpha(a)].$$

Similarly,

$$|L(P_1) - L(P_2)| \leq 2\epsilon[\alpha(b) - \alpha(a)].$$

Since ϵ may be chosen arbitrarily small, it follows that if $\{P_n\}$ is any sequence of partitions with $|P_n| \to 0$, the sequences $\{U(P_n)\}$ and $\{L(P_n)\}$ are Cauchy, and hence converge. By Eq. (1) they converge to the same limit. The observation that $U(P_1)$ and $U(P_2)$ must be close to each other for small $|P_1|$ and $|P_2|$ implies that the limit is the same regardless of the particular sequence $\{P_n\}$. ■

Note that by Eq. (3), $L(P_n) \leq S(P_n) \leq U(P_n)$ for all n; hence, $S(P) \to \int_a^b f \, d\alpha$ as $|P| \to 0$. This may be used as the basis of a generalization of the Riemann–Stieltjes integral in which the monotonicity requirement on α is removed.

6.1.3 Definition

Let f: $[a, b] \to R$, α: $[a, b] \to R$, and assume f is bounded. We say that $\int_a^b f \, d\alpha$ exists if $S(P, f, \alpha)$ approaches a finite limit A as $|P| \to 0$. In other words, given $\epsilon > 0$ there is a $\delta > 0$ such that whenever $|P| < \delta$ (regardless of the choice of the t_k), we have $|S(P, f, \alpha) - A| < \epsilon$. Equivalently, if $\{P_n\}$ is any sequence of partitions with $|P_n| \to 0$, we have $S(P_n, f, \alpha) \to A$.

We know that if f is continuous and α is increasing, then $\int_a^b f \, d\alpha$ exists. The case in which f is continuous and α is decreasing is handled similarly (if we like, we can multiply α by -1 so that $-\alpha$ will be increasing).

Problems for Section 6.1

1. Give an example in which f is Riemann integrable on $[a, b]$ and α is increasing, but $\int_a^b f \, d\alpha$ does not exist.
2. Let $f(x) = 1, x$ rational; $f(x) = 0, x$ irrational. Show that f is not Riemann integrable on [0,1].
3. Give an example of a function f that is not Riemann integrable but whose absolute value $|f|$ is Riemann integrable.

4. Assume f bounded and α increasing on $[a, b]$. By Eq. (3) if $U(P, f, \alpha)$ and $L(P, f, \alpha)$ approach a common finite limit $\int_a^b f\, d\alpha$ as $|P| \to 0$, then $S(P, f, \alpha) \to \int_a^b f\, d\alpha$. Prove that, conversely, if $S(P, f, \alpha) \to \int_a^b f\, d\alpha$ (finite), then $U(P, f, \alpha)$ and $L(P, f, \alpha)$ also approach $\int_a^b f\, d\alpha$. Thus, the characterizations of the integral via upper and lower sums and via Riemann–Stieltjes sums are equivalent.

6.2 PROPERTIES OF THE INTEGRAL

A number of basic properties of the integral follow directly from the definition. To save space, we adopt the notation $f \in R(\alpha)$ to indicate that $\int_a^b f\, d\alpha$ exists.

6.2.1 THEOREM

(a) If $f_1, f_2 \in R(\alpha)$, then $f_1 + f_2 \in R(\alpha)$ and

$$\int_a^b (f_1 + f_2)\, d\alpha = \int_a^b f_1\, d\alpha + \int_a^b f_2\, d\alpha.$$

(b) If $f \in R(\alpha)$ and c is a constant, then $cf \in R(\alpha)$ and

$$\int_a^b cf\, d\alpha = c \int_a^b f\, d\alpha.$$

Also, $f \in R(c\alpha)$ and $\int_a^b f\, d(c\alpha) = c \int_a^b f\, d\alpha$.
(c) If $f \in R(\alpha_1)$, and $f \in R(\alpha_2)$, then $f \in R(\alpha_1 + \alpha_2)$ and

$$\int_a^b f\, d(\alpha_1 + \alpha_2) = \int_a^b f\, d\alpha_1 + \int_\alpha^b f\, d\alpha_2.$$

For the remaining statements, assume α is increasing.
(d) If $f_1, f_2 \in R(\alpha)$ and $f_1 \le f_2$ on $[a, b]$, then

$$\int_a^b f_1\, d\alpha \le \int_a^b f_2\, d\alpha.$$

(e) If $f \in R(\alpha)$, then $|f| \in R(\alpha)$ and

$$\left|\int_a^b f\,d\alpha\right| \leq \int_a^b |f|\,d\alpha.$$

Thus, if $|f| \leq M$ on $[a,b]$, we have

$$\left|\int_a^b f\,d\alpha\right| \leq M\,[\alpha(b) - \alpha(a)].$$

(f) Assume $a < c < b$. If $f \in R(\alpha)$ on $[a,c]$ and $f \in R(\alpha)$ on $[c,b]$, then $f \in R(\alpha)$ on $[a,b]$ and

$$\int_a^b f\,d\alpha = \int_a^c f\,d\alpha + \int_c^b f\,d\alpha.$$

(g) If f is Riemann integrable on $[a,b]$ and $g = f$ except at finitely many points $y_1, \ldots, y_r$, then g is Riemann integrable on $[a,b]$ and $\int_a^b g(x)\,dx = \int_a^b f(x)\,dx$.
(h) If f is piecewise continuous on $[a,b]$, that is, f has only finitely many discontinuities on $[a,b]$, all removable or jumps, then f is Riemann integrable on $[a,b]$.

Proof

(a)

$$\begin{aligned} S(P, f_1 + f_2, \alpha) &= \sum_{k=1}^{n} [f_1(t_k) + f_2(t_k)]\,\Delta\alpha_k \\ &= \sum_{k=1}^{n} f_1(t_k)\,\Delta\alpha_k + \sum_{k=1}^{n} f_2(t_k)\,\Delta\alpha_k \\ &\to \int_a^b f_1\,d\alpha + \int_a^b f_2\,d\alpha \qquad \text{as} \quad |P| \to 0. \end{aligned}$$

(b)

$$S(P,cf,\alpha) = S(P,f,c\alpha) = c\sum_{k=1}^{n} f(t_k)\,\Delta\alpha_k$$
$$\to c\int_a^b f\,d\alpha \qquad \text{as} \quad |P| \to 0.$$

(c)

$$S(P,f,\alpha_1+\alpha_2) =$$
$$\sum_{k=1}^{n} f(t_k)[\alpha_1(x_k)+\alpha_2(x_k)-(\alpha_1(x_{k-1})+\alpha_2(x_{k-1}))]$$
$$\to \int_a^b f\,d\alpha_1 + \int_a^b f\,d\alpha_2.$$

(d)

$$S(P,f_1,\alpha) = \sum_{k=1}^{n} f_1(t_k)\,\Delta\alpha_k \le \sum_{k=1}^{n} f_2(t_k)\,\Delta\alpha_k = S(P,f_2,\alpha)$$

(since $f_1 \le f_2$ and α is increasing, so $\Delta\alpha_k \ge 0$). Let $|P| \to 0$ to obtain the desired result.

(e)

$$S(P,f,\alpha) = \sum_{k=1}^{n} f(t_k)\,\Delta\alpha_k;$$

hence,

$$|S(P,f,\alpha)| \le \sum_{k=1}^{n} |f(t_k)|\,\Delta\alpha_k = S(P,|f|,\alpha).$$

Let $|P| \to 0$ to get $|\int_a^b f\,d\alpha| \le \int_a^b |f|\,d\alpha$, provided $|f| \in R(\alpha)$. But $|f(y)| = |f(y)-f(x)+f(x)| \le |f(y)-f(x)| + |f(x)|$, and similarly $|f(x)| \le |f(y) - f(x)| + |f(y)|$. Consequently, $||f(y)| - |f(x)|| \le |f(y) - f(x)|$, and it follows that replacing f by $|f|$ decreases $|U(P) - L(P)|$. (Let $|f(y)|$ approximate $\sup |f|$, and let $|f(x)|$ approximate $\inf |f|$.) Consequently, $|f| \in R(\alpha)$. (See Problem 1.)

(f) We may assume that either f or α is continuous at c; otherwise, $\int_a^c f \, d\alpha$ or $\int_c^b f \, d\alpha$ fails to exist (cf. Section 6.1, Problem 1). Given any sequence of partitions P of $[a, b]$ with $|P| \to 0$, and given $\epsilon > 0$, we wish to refine P by adding the point c, producing a new partition Q with $|U(P, f, \alpha) - U(Q, f, \alpha)| < \epsilon$ and $|L(P, f, \alpha) - L(Q, f, \alpha)| < \epsilon$. The discussion following Eq. (3) of 6.1.1 shows that this can be accomplished for sufficiently small $|P|$ (in that discussion y is replaced by c). Since Q is formed by a partition of $[a, c]$ followed by a partition of $[c, b]$, $U(Q, f, \alpha)$ and $L(Q, f, \alpha) \to \int_a^c f \, d\alpha + \int_c^b f \, d\alpha$ as $|P|$, hence $|Q|$, approaches 0. It follows that for sufficiently small $|P|$, $U(P, f, \alpha)$ and $L(P, f, \alpha)$ differ from $\int_a^c f \, d\alpha + \int_c^b f \, d\alpha$ by less than ϵ. Since $\epsilon > 0$ is arbitrary, the proof is complete.

(g) If $S(P, f) = \sum_{k=1}^n f(t_k)[x_k - x_{k-1}]$, and $S(P, g) = \sum_{k=1}^n g(t_k)[x_k - x_{k-1}]$, then

$$
\begin{aligned}
|S(P, f) - S(P, g)| &\le \sum_{k=1}^{n} |f(t_k) - g(t_k)|[x_k - x_{k-1}] \\
&\le \sum_{j=1}^{r} |f(y_j) - g(y_j)||P| \\
&\to 0 \qquad \text{as} \quad |P| \to 0.
\end{aligned}
$$

(h) The interval $[a, b]$ can be written as $[y_0, y_1] \cup [y_1, y_2] \cup \cdots \cup [y_{r-1}, y_r]$, where the discontinuities of f occur at the y_i. By (g), f is Riemann integrable on each subinterval, since it becomes continuous when changed at the end points. By (f), f is Riemann integrable on $[a, b]$. ■

It follows from Theorem 6.2.1(b) and (c) that if f is continuous and α is the difference of two increasing functions, then $\int_a^b f \, d\alpha$ exists. We will see in the next section that if α has a continuous derivative α' on $[a, b]$ then α can be expressed as $\alpha_1 - \alpha_2$, where α_1 and α_2 are increasing; it follows that $f \in R(\alpha)$. (In Chapter 5 we omitted any discussion of the "one-sided" derivative at the end points of a closed interval, but the definition is natural:

$$\alpha'(a) = \lim_{\substack{h\to 0\\ h>0}} \frac{\alpha(a+h)-\alpha(a)}{h}, \qquad \alpha'(b) = \lim_{\substack{h\to 0\\ h<0}} \frac{\alpha(b+h)-\alpha(b)}{h}.\Bigg)$$

These comments allow us to show that a Riemann–Stieltjes integral may often be reduced to a Riemann integral, as follows.

6.2.2 THEOREM. *If f is continuous on $[a, b]$ and α has a continuous derivative on $[a, b]$, then*

$$\int_a^b f(x)\,d\alpha(x) = \int_a^b f(x)\alpha'(x)\,dx.$$

Proof. Let P: $a = x_0 < x_1 < \cdots < x_n = b$ be a partition of $[a, b]$. By the Mean Value Theorem 5.1.4, $\alpha(x_k) - \alpha(x_{k-1}) = \alpha'(t_k)(x_k - x_{k-1})$ for some $t_k \in (x_{k-1}, x_k)$. The points $t_1, \ldots, t_n$ yield the following Riemann–Stieltjes sum:

$$S(P, f, \alpha) = \sum_{k=1}^{n} f(t_k)\alpha'(t_k)(x_k - x_{k-1}).$$

Let $|P| \to 0$ to obtain

$$\int_a^b f\,d\alpha = \int_a^b f(x)\alpha'(x)\,dx,$$

as desired. ■

We are now in a position to evaluate Riemann–Stieltjes integrals in practical cases. In the usual physical situation, $\alpha = \alpha_1 + \alpha_2$ where α_1 has a continuous derivative and α_2 is a *jump function* having jumps of size c_n at the points x_n, $n = 1, 2, \ldots N$; α_2 is constant between jumps (see Fig. 6.2.1). In our discussion we assume α_2 is increasing, but the evaluation formula to be obtained is valid for any jump function α_2, since α_2 is expressible as the difference of two increasing jump functions.

Figure 6.2.1 Jump Function

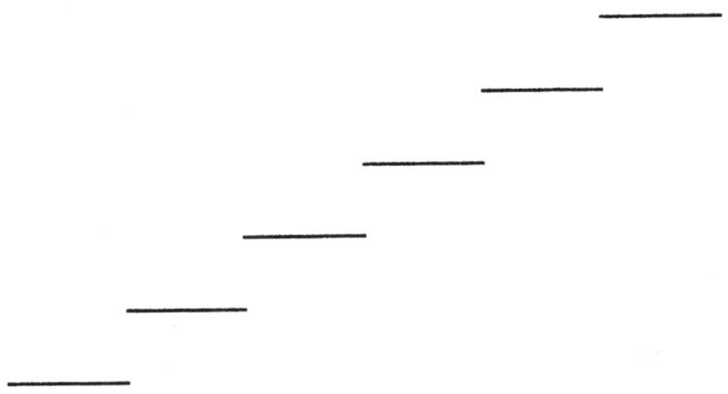

Consider the behavior of α_2 at a jump of size $\Delta\alpha_2(c)$ at $x = c$ (see Fig. 6.2.2). Assuming f continuous, we know that $S(P, f, \alpha_2) \to \int_a^b f \, d\alpha_2$ as $|P| \to 0$, so we are free to choose any convenient sequences of P's. If we choose P so that $[c-h, c+h]$, $h > 0$, is one of the subintervals of P, we obtain a contribution of the form

$$f(t)[\alpha_2(c+h) - \alpha_2(c-h)] \to f(c)\,\Delta\alpha_2(c)$$

as $h \to 0$, by continuity of f. In general, a jump of size c_n at x_n contributes $f(x_n)c_n$ to $\int_a^b f \, d\alpha_2$. We have arrived at the following result.

6.2.3 Evaluation Formula

Let f be continuous on $[a, b]$, and let $\alpha = \alpha_1 + \alpha_2$, where α_1' is continuous on $[a, b]$ and α_2 is a jump function. Assume that the jumps of α_2 occur at $x_1, \ldots, x_N$, and let $c_n = \alpha_2(x_n^+) - \alpha_2(x_n^-)$, $1 \le n \le N$; notice that the actual value of $\alpha_2(x)$ at $x = c_n$ is irrelevant. Then

$$\int_a^b f(x)\, d\alpha(x) = \int_a^b f(x)\alpha_1'(x)\, dx + \sum_{n=1}^{N} f(x_n)c_n.$$

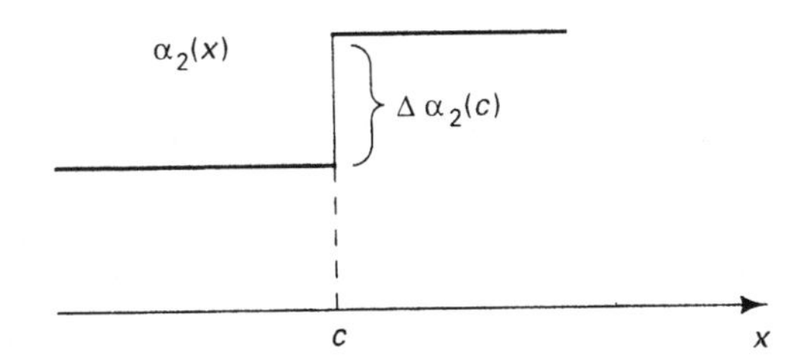

Figure 6.2.2 Behavior at a Jump

The evaluation formula may be extended to the case where f has a finite number of simple discontinuities as long as f and α are not discontinuous at the same point; when simultaneous discontinuities occur, the Riemann–Stieltjes integral does not exist (see Problem 1 of Section 6.1).

A basic application of Riemann–Stieltjes integrals occurs in probability. The function α is the "distribution function" of a "random variable" X. Intuitively, if α has a jump of size c_n at x_n, then the probability that X will take the value x_n in a particular performance of the random experiment is c_n. If $\sum_{n=1}^{N} c_n < 1$, then X has a "continuous component" (as well as a discrete component represented by the set of probabilities c_n). The term $\alpha_1'(x)\ dx$ represents the probability that X will take a value between x and $x + dx$, assuming none of the x_i, $1 \leq i \leq N$, belong to $[x, x + dx]$. The evaluation formula then gives the average value of $f(X)$. For example, if $f(x) = x^2$, then the formula gives the average value of the square of the random variable.

Another basic application is to line integrals. For example, if C is a curve in the plane, parametrized by $x = x(t)$, $y = y(t)$, $a \leq t \leq b$, let s be arc length along the curve, so that $s(t_0)$ is the length of the portion of the curve given by $(x(t), y(t))$, $a \leq t \leq t_0$. The line integral $\int_C f(x,y)\ ds$ is given by the Riemann-Stieltjes integral

$$\int_a^b f(x(t), y(t))\ ds(t).$$

In particular, if f is the density of a curved rod, then $\int_C f(x,y)\ ds$ is the total mass.

In the evaluation formula, the integral $\int_a^b f(x)\alpha_1'(x)\ dx$ is evaluated by standard calculus techniques. The following result justifies such a computation.

6.2.4 Fundamental Theorem of Calculus

Let f be a continuous real-valued function on $[a, b]$.

(a) *If $F(x) = \int_a^x f(t)\ dt$, $a \leq x \leq b$, then $F' = f$ on $[a, b]$.*

(b) *If* $G' = f$ *on* $[a,b]$, *then* $\int_a^b f(x)\,dx = G(b) - G(a)$.

Proof

(a) Let $h > 0$ (the argument for $h < 0$ is similar). Then

$$\begin{aligned}\left|\frac{F(x+h)-F(x)}{h} - f(x)\right| &= \left|\frac{1}{h}\int_x^{x+h} [f(t)-f(x)]\,dt\right| \\ &\le \frac{1}{h}\int_x^{x+h} |f(t)-f(x)|\,dt \\ &\le \max_{x \le t \le x+h} |f(t)-f(x)| \\ &\to 0 \quad \text{as} \quad h \to 0\end{aligned}$$

by continuity of f.
(b) By (a), $(d/dx)(G(x)-F(x)) = f(x) - f(x) = 0$ on $[a,b]$, and hence $G - F$ is constant on $[a,b]$ (use the Mean Value Theorem). Thus, $G(b) - G(a) = F(b) - F(a) = \int_a^b f(x)\,dx$. ■

For an intuitive view of Theorem 6.2.4(a), represent F by the cumulative area A under f between a and x. If we change x by a "small" amount dx, the cumulative area changes by (approximately) $f(x)\,dx$ (see Fig. 6.2.3). Thus $dA = f(x)\,dx$, or $dA/dx = f(x)$, which says that $F' = f$.

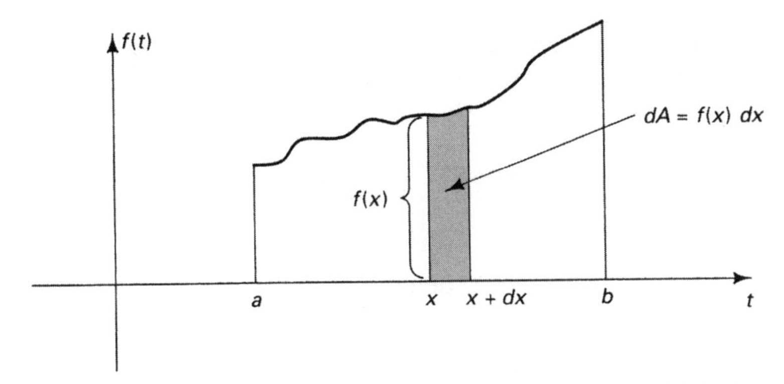

Figure 6.2.3 Intuitive View of the Fundamental Theorem of Calculus

Problems for Section 6.2

1. If α is increasing and f is bounded on $[a,b]$, show that $f \in R(\alpha)$ iff $U(P,f,\alpha) - L(P,f,\alpha) \to 0$ as $|P| \to 0$.
2. Assume α increasing and f bounded on $[a,b]$. If $a < c < b$ and $f \in R(\alpha)$ on $[a,b]$, show that $f \in R(\alpha)$ on both $[a,c]$ and $[c,b]$.
3. Let $f(x) = x^2$, and define α as follows:

$$\alpha(x) = \begin{cases} 0, & x < 2, \\ \frac{2}{3}x - 1, & 2 \leq x < 3, \\ 1, & x \geq 3. \end{cases}$$

   Evaluate $\int_0^3 f\, d\alpha$.
4. Evaluate $\int_0^n x\, d\alpha(x)$ where $\alpha(x) = [x] =$ the largest integer $\leq x$.

6.3 FUNCTIONS OF BOUNDED VARIATION

We have seen that if f is continuous and α is the difference of two increasing functions, then $f \in R(\alpha)$. We now look at the question of when a function α can be expressed as $\alpha_1 - \alpha_2$, where α_1 and α_2 are increasing.

6.3.1 Definitions and Comments

Let $f\colon [a,b] \to R$, and let $P\colon a = x_0 < x_1 < \cdots < x_n = b$ be a partition of $[a,b]$. Define

$$V(f,P) = \sum_{i=1}^{n} |f(x_i) - f(x_{i-1})|.$$

The *variation* of f on $[a,b]$ is defined by

$$V(f) = \sup_P V(f,P).$$

If we wish to emphasize the interval $[a, b]$ on which the computation is carried out, we write $V(f; [a, b])$ instead of simply $V(f)$.

We say that f is of *bounded variation* (on $[a, b]$) if $V(f) < \infty$.

It follows from the definitions that a monotone function is of bounded variation, with $V(f) = |f(b) - f(a)|$. Also, for arbitrary f we have $V(-f) = V(f)$. A brief computation shows that

$$V(f + g) \le V(f) + V(g).$$

To see this, note that $V(f+g, P) \le V(f, P) + V(g, P)$ by the triangle inequality; hence, $V(f + g, P) \le V(f) + V(g)$ for any P. Take the sup over P to obtain the desired result. Observe also that since $V(-g) = V(g)$, we have $V(f - g) \le V(f) + V(g)$. It follows that if $\alpha = f - g$, where f and g are increasing, then α is of bounded variation.

We adopt the notation $f \in BV$ to indicate that f is of bounded variation. The main result is that if $\alpha \in BV$ then α can be expressed as the difference of two increasing functions. It follows that if f is continuous and $\alpha \in BV$, then $\int_a^b f\, d\alpha$ exists. Before proving this, we examine some basic properties and examples.

6.3.2 THEOREM

(a) If $f \in BV$, then f is bounded.
(b) If f, $g \in BV$ with $|f| \le A$ and $|g| \le B$ on $[a, b]$, then $V(fg) \le AV(g) + BV(f)$; hence, $fg \in BV$.
(c) If f' exists and is bounded on $[a, b]$, then $f \in BV$.
(d) If $f(x) = x \sin(1/x)$, $0 \le x \le b$, $f(0) = 0$, then f is continuous but not of bounded variation.
(e) If $a < c < b$, then $V(f; [a, b]) = V(f; [a, c]) + V(f; [c, b])$.

Proof

(a) If $a < x < b$, we have

$$|f(x) - f(a)| \le |f(x) - f(a)| + |f(b) - f(x)| \le V(f) < \infty.$$

(b) A typical term in the computation of $V(fg, P)$ is

$$|f(x_i)g(x_i) - f(x_{i-1})g(x_{i-1})| \le |f(x_i)||g(x_i) - g(x_{i-1})| + |g(x_{i-1})||f(x_i) - f(x_{i-1})|.$$

Since $|f(x_i)| \le A$ and $|g(x_{i-1})| \le B$, it follows that

$$\begin{aligned} V(fg, P) &\le AV(g, P) + BV(f, P) \\ &\le AV(g) + BV(f) \qquad \text{for all} \quad P. \end{aligned}$$

Take the sup over P to obtain the desired result.

(c) By the Mean Value Theorem,

$$\begin{aligned} |f(x_i) - f(x_{i-1})| &= |f'(t_i)||x_i - x_{i-1}| \\ &\qquad \text{for some } t_i \in (x_{i-1}, x_i) \\ &\le M|x_i - x_{i-1}| \qquad \text{if} \quad |f'| \le M. \end{aligned}$$

Thus, $V(f) \le M(b-a) < \infty$.

(d) Let $y_n = 2/n\pi$, $n = 1, 2, \ldots$. Then

$$f(y_1) = \frac{2}{\pi}, \quad f(y_2) = 0, \quad f(y_3) = \frac{-2}{3\pi}, \quad f(y_4) = 0,$$
$$f(y_5) = \frac{2}{5\pi}, \ldots, \quad f(y_{2n}) = 0, \quad f(y_{2n+1}) = \pm\frac{2}{(2n+1)\pi}.$$

Let P_n be the partition formed by $0, y_{2n+1}, y_{2n}, \ldots, y_1$. Then

$$\begin{aligned} V(f, P_n) &= \frac{2}{\pi}\left(1 + \frac{1}{3} + \frac{1}{5} + \frac{1}{7} + \cdots + \frac{1}{2n+1}\right) \\ &> \frac{2}{\pi}\left(\frac{1}{4} + \frac{1}{6} + \frac{1}{8} + \cdots + \frac{1}{2n+2}\right) \\ &= \frac{1}{\pi}\left(\frac{1}{2} + \frac{1}{3} + \frac{1}{4} + \cdots + \frac{1}{n+1}\right) \\ &\to \infty \qquad \text{as} \quad n \to \infty. \end{aligned}$$

(e) If P is any partition of $[a, c]$ and Q is any partition of $[c, b]$, we have $V(f, P) + V(f, Q) \le V(f; [a, b])$, so $V(f; [a, c]) +$

$V(f;[c,b]) \le V(f;[a,b])$. If P_0 is any partition of $[a,b]$, we can refine P_0 if necessary to obtain a partition P' of $[a,c]$ followed by a partition Q' of $[c,b]$. The refining process can only increase the variation, by the triangle inequality. Thus,

$$\begin{aligned} V(f,P_0) &\le V(f,P') + V(f,Q') \\ &\le V(f;[a,c]) + V(f;[c,b]). \end{aligned}$$

Take the sup over P_0 to finish the argument. ■

We now prove the main result on functions of bounded variation.

6.3.3 THEOREM. *If f is of bounded variation on $[a,b]$, define*

$$F(x) = V(f;[a,x]), \quad a \le x \le b, \qquad G(x) = F(x) - f(x).$$

Then F and G are increasing on $[a,b]$, so f is expressible as the difference of two increasing functions.

Proof. Note that F "follows" f, reflecting decreasing portions about the horizontal to obtain increasing portions; see Fig. 6.3.1. It follows from Theorem 6.3.2(e) that F is increasing. To show that G is increasing, let $x < y$. Then

$$\begin{aligned} G(y) - G(x) &= F(y) - F(x) - (f(y) - f(x)) \\ &= V(f;[a,y]) - V(f;[a,x]) - (f(y) - f(x)) \\ &= V(f;[x,y]) - (f(y) - f(x)) \quad \text{by Theorem 6.3.2(e)} \\ &\ge V(f;[x,y]) - |f(y) - f(x)| \\ &\ge 0 \qquad \text{by definition of variation.} \end{aligned}$$ ■

If f is continuous on $[a,b]$, then so are F and G; see the problems.

Problems for Section 6.3

The purpose of this problem set is to show that, in Theorem 6.3.3, if f is continuous on $[a,b]$ so are F and G.

1. Let $c \in [a,b)$, $\epsilon > 0$. Show that there is a $\delta > 0$ and a partition P: $c = x_0 < x_1 < \cdots < x_n = b$ such that $0 < x_1 - x_0 <$

Figure 6.3.1 Expressing a Function of Bounded Variation as the Difference of Two Increasing Functions

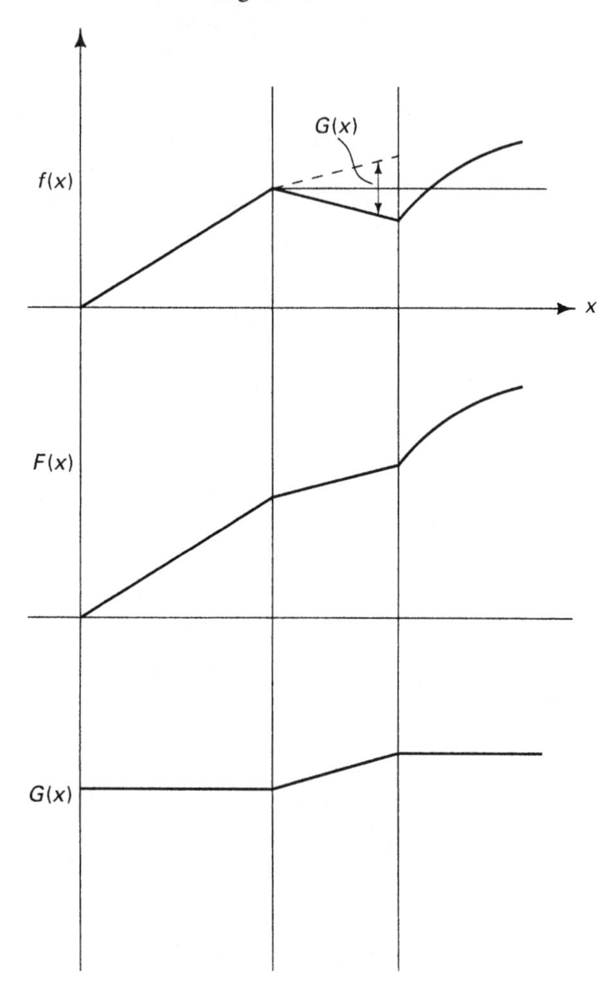

δ, $|f(x_1) - f(x_0)| < \epsilon/2$, and $\sum_{k=1}^{n} |f(x_k) - f(x_{k-1})| > V(f; [c, b]) - \epsilon/2$.

2. Show that $F(b) - F(c) - \epsilon/2 < \epsilon/2 + F(b) - F(x_1)$, and conclude that F is *right-continuous* at c; i.e.; $\lim_{x \to c;\, x > c} F(x) = F(c)$.

3. Let $a < c \leq b$, $\epsilon > 0$. Show that there is a $\delta > 0$ and a partition P : $a = x_0 < x_1 < \cdots < x_{n-1} < x_n = c$ such that $|f(x_n) - f(x_{n-1})| < \epsilon/2$, $0 < x_n - x_{n-1} < \delta$, and $\sum_{k=1}^{n} |f(x_k) - f(x_{k-1})| > V(f; [a, c]) - \epsilon/2$.

4. Show that $F(c) - F(a) - \epsilon/2 < \epsilon/2 + F(x_{n-1}) - F(a)$, and conclude that F is *left-continuous* at c; i.e. $\lim_{x \to c;\, x<c} F(x) = F(c)$. It follows that F, hence G, is continuous on $[a, b]$.

6.4 SOME USEFUL INTEGRATION THEOREMS

We close this chapter with some results that are often useful in computations.

6.4.1 Integration by Parts

If $\int_a^b f\, d\alpha$ exists, then so does $\int_a^b \alpha\, df$, and

$$\int_a^b f\, d\alpha + \int_a^b \alpha\, df = f(b)\alpha(b) - f(a)\alpha(a).$$

Proof. Consider any Riemann–Stieltjes sum for $\int_a^b \alpha\, df$, say

$$S(P, \alpha, f) = \sum_{k=1}^{n} \alpha(t_k)[f(x_k) - f(x_{k-1})].$$

Now

$$\begin{aligned}\sum_{k=1}^{n} f(x_k)\alpha(x_k) - \sum_{k=1}^{n} f(x_{k-1})\alpha(x_{k-1}) &= f(x_n)\alpha(x_n) - f(x_0)\alpha(x_0)\\ &= f(b)\alpha(b) - f(a)\alpha(a).\end{aligned}$$

Call this last line A. Then

$$\begin{aligned}A - S(P, \alpha, f) = &\sum_{k=1}^{n} f(x_k)[\alpha(x_k) - \alpha(t_k)]\\ &+ \sum_{k=1}^{n} f(x_{k-1})[\alpha(t_k) - \alpha(x_{k-1})].\end{aligned}$$

Since $x_{k-1} \le t_k \le x_k$, this is a Riemann–Stieltjes sum for $\int_a^b f\, d\alpha$ corresponding to a partition P' refining P. It follows that as $|P| \to 0$,

$A - S(P, \alpha, f) \to \int_a^b f\, d\alpha$. Thus, $S(P, \alpha, f) \to A - \int_a^b f\, d\alpha$, as desired. ■

6.4.2 Change of Variable Formula

Let f and h be continuous functions from $[a, b]$ to R, and assume h strictly increasing. Let α be the inverse function of h; in other words, if $y = h(x)$, then $x = \alpha(y)$, $a \le x \le b$, $h(a) \le y \le h(b)$. Then

$$\int_a^b f(x)\, dx = \int_{h(a)}^{h(b)} f(\alpha(y))\, d\alpha(y).$$

Proof. Let P: $a = x_0 < x_1 < \cdots < x_n = b$, Q: $h(a) = y_0 < y_1 < \cdots < y_n = h(b)$, where $y_k = h(x_k)$, $0 \le k \le n$. (Since h is strictly increasing, so is α.) Then

$$(1) \quad \sum_{k=1}^{n} f(x_k)(x_k - x_{k-1}) = \sum_{k=1}^{n} f(\alpha(y_k))(\alpha(y_k) - \alpha(y_{k-1})).$$

Since h and α are continuous functions defined on closed, bounded (therefore compact) intervals, they are uniformly continuous. (Continuity of α follows from Section 4.2, Problem 5). Thus, if $\epsilon > 0$, there is a $\delta > 0$ such that if $|x_k - x_{k-1}| < \delta$ then $|y_k - y_{k-1}| < \epsilon$. It follows that if $|P| \to 0$ then $|Q| \to 0$ (and conversely). Now $f \circ \alpha$ is a composition of continuous functions and is therefore continuous. Since α is monotone, $\int_{h(a)}^{h(b)} f(\alpha(y))\, d\alpha(y)$ exists. Thus, if we let $|P| \to 0$ in (1), we obtain $\int_a^b f(x)\, dx = \int_{h(a)}^{h(b)} f(\alpha(y))\, d\alpha(y)$, as desired. ■

6.4.3 Mean Value Theorem for Integrals

If f is continuous and α is increasing on $[a, b]$, there is a point $x_0 \in [a, b]$ such that $\int_a^b f\, d\alpha = f(x_0)[\alpha(b) - \alpha(a)]$.

Proof. Let $M = \sup\{f(t) : a \le t \le b\}$ and $m = \inf\{f(t) : a \le t \le b\}$. Then $m(\alpha(b) - \alpha(a)) \le \int_a^b f\, d\alpha \le M(\alpha(b) - \alpha(a))$; hence,

$$m \le \frac{1}{\alpha(b) - \alpha(a)} \int_a^b f\, d\alpha \le M.$$

(Note that if $\alpha(a) = \alpha(b)$ then α is constant on $[a, b]$ and the result is trivial.) Thus $(\alpha(b)-\alpha(a))^{-1} \int_a^b f\, d\alpha = c$ for some $c \in [m, M]$. Now if, say, $f(x_1) = m$ and $f(x_2) = M$, then by the Intermediate Value Theorem there is a point x_0 between x_1 and x_2 such that $f(x_0) = c$. We conclude that $\int_a^b f\, d\alpha = f(x_0)[\alpha(b) - \alpha(a)]$. ■

6.4.4 Upper Bounds on Integrals

Assume $\int_a^b f\, d\alpha$ exists, and let $M = \sup\{|f(x)| : a \leq x \leq b\}$. Then

$$\left| \int_a^b f\, d\alpha \right| \leq M V(\alpha),$$

where $V(\alpha)$ is the variation of α on $[a, b]$. Thus, if α is increasing, we have

$$\left| \int_a^b f\, d\alpha \right| \leq M[\alpha(b) - \alpha(a)].$$

If $\alpha(x) = x$, we obtain a bound on the Riemann integral:

$$\left| \int_a^b f(x)\, dx \right| \leq M(b - a).$$

Proof. Consider an arbitrary Riemann–Stieltjes sum $S(P, f, \alpha) = \sum_{k=1}^{n} f(t_k)\, \Delta\alpha_k$. We have

$$\begin{aligned} |S(P, f, \alpha)| &\leq \sum_{k=1}^{n} |f(t_k)||\Delta\alpha_k| \\ &\leq M \sum_{k=1}^{n} |\Delta\alpha_k| \\ &\leq M V(\alpha). \end{aligned}$$ ■

6.4.5 Improper Integrals

An integral in which one or both limits of integration are infinite is called *improper;* it is defined as a limit of ordinary integrals. For example,

$$\int_{-\infty}^{\infty} f\,d\alpha = \lim_{\substack{a\to-\infty\\ b\to\infty}} \int_a^b f\,d\alpha \qquad \text{if the limit exists.}$$

In many practical cases, integrals of this type are evaluated in the usual way; for example, $\int_0^\infty e^{-x}\,dx = -e^{-x}|_0^\infty = 1$. For a formal justification, note that $\int_0^b e^{-x}\,dx = 1 - e^{-b} \to 1$ as $b \to \infty$.

An integral in which the function f is unbounded is also referred to as improper. The theory of improper integrals is best handled within the domain of measure theory. As an example of what might happen, consider the function of Fig. 6.4.1. Here, f is bounded and continuous everywhere, but $\int_0^\infty f(x)\,dx$ does not exist because $\int_0^{2n} f(x)\,dx$ oscillates between 0 and 1 for $n = 0, 1, 2, \ldots$. In general, it turns out that if $\int_{-\infty}^{\infty} |f(x)|\,dx$ exists and is finite—that is, $\int_a^b |f(x)|\,dx$ approaches a finite limit as $a \to -\infty$ and $b \to \infty$—then $\int_{-\infty}^{\infty} f(x)\,dx$ exists and is finite also.

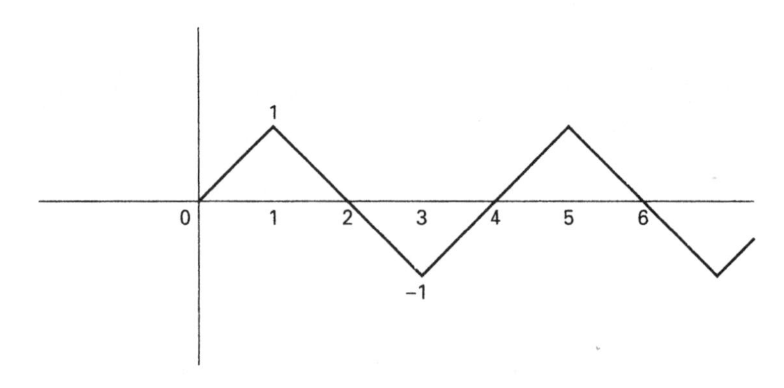

Figure 6.4.1 Improper Riemann Integral

Problems for Section 6.4

1. If f and g are continuous and α increasing on $[a,b]$, show that

$$\left|\int_b^a fg\, d\alpha\right|^2 \le \int_a^b f^2\, d\alpha \int_a^b g^2\, d\alpha.$$

This is the *Cauchy–Schwarz inequality for integrals*; the analogous result for sums is given in Section 3.2, Problem 3.

2. Prove the *integral test* for convergence of series: if f is continuous, nonnegative, and decreasing, show that $\int_1^\infty f(x)\, dx$ is finite (that is, $\int_1^b f(x)\, dx$ approaches a finite limit as $b \to \infty$) if and only if $\sum_{n=1}^\infty f(n) < \infty$.

3. Give an example of a function f for which the improper Riemann integral $\int_1^\infty f(x)\, dx$ is finite, but $\int_0^\infty |f(x)|\, dx = \infty$.

4. Give an example of an unbounded function f on $(0,1]$ such that the improper integral $\int_0^1 f(x)\, dx$ is
   (a) finite.
   (b) infinite.

REVIEW PROBLEMS FOR CHAPTER 6

1. Evaluate $\int_1^4 f\, d\alpha$, where $f(x) = 1/x$ and

$$\alpha(x) = \begin{cases} x^2 & \text{if } 1 \le x < 2, \\ 3 + 3x & \text{if } 2 \le x < 3, \\ x^3 & \text{if } 3 \le x \le 4. \end{cases}$$

2. True or false, and explain briefly.
   (a) If f is piecewise continuous on $[a,b]$ and α is increasing on $[a,b]$, then $\int_a^b f\, d\alpha$ exists.
   (b) Let f be continuous and α increasing on $[a,b]$. If Q is a refinement of the partition P, then $L(Q,f,\alpha) \ge L(P,f,\alpha)$.

(c) If f is increasing and α is continuous on $[a, b]$, then $\int_a^b f \, d\alpha$ exists.

3. Let f be continuous on R, and let g and h be differentiable on R. Show that

$$\frac{d}{dx} \int_{g(x)}^{h(x)} f(t)\, dt = f(h(x))h'(x) - f(g(x))g'(x).$$

4. Give an example of a function $f\colon [0,1] \to R$ such that f is not Riemann integrable on $[0,1]$ but f^2 is Riemann integrable on $[0,1]$.

5. Let

$$\alpha(x) = \begin{cases} 2|x| & \text{if } x < 3, \\ 10 & \text{if } x \geq 3. \end{cases}$$

If $f(x) = 3x$, evaluate $\int_{-1}^{5} f \, d\alpha$.

6. Let f be a continuous, real-valued function on $[a, b]$, and let $F(x) = \int_a^x f(t)\, dt$. Show that F is of bounded variation on $[a, b]$.

7. A student evaluates the integral $\int_{-1}^{1} x^2 \, dx$ as follows: Make the change of variable $y = x^2$ to get

$$\int_1^1 \quad \text{(who cares)} \quad dy = 0$$

because $\int_a^a \ldots$ is always 0.

But

$$\int_{-1}^{1} x^2 \, dx = \frac{x^3}{3}\Big|_{-1}^{1} = \frac{2}{3}.$$

Where did the student go wrong?

7

UNIFORM CONVERGENCE AND APPLICATIONS

7.1 POINTWISE AND UNIFORM CONVERGENCE

We have encountered several operations that are performed on functions or sequences of functions, including differentiation, integration, and the taking of limits. Often, two operations are performed in sequence, and it may make a difference in which order the operations are carried out. We illustrate this point with some examples.

7.1.1 Examples of Invalid Interchange of Operations

Let

(a)
$$f_n(x) = \begin{cases} 0, & x < 0, \\ nx, & 0 \le x < 1/n, \\ 1, & x \ge 1/n. \end{cases}$$

For each fixed x, $f_n(x) \to f(x)$ as $n \to \infty$, where $f(x) = 0$ for $x \le 0$, and $f(x) = 1$ for $x > 0$ (see Fig. 7.1.1). We consider the operations (on $f_n(x)$) of taking the limit as $x \to 0$ and taking the limit as $n \to \infty$.

Figure 7.1.1 Example 7.1.1(a)

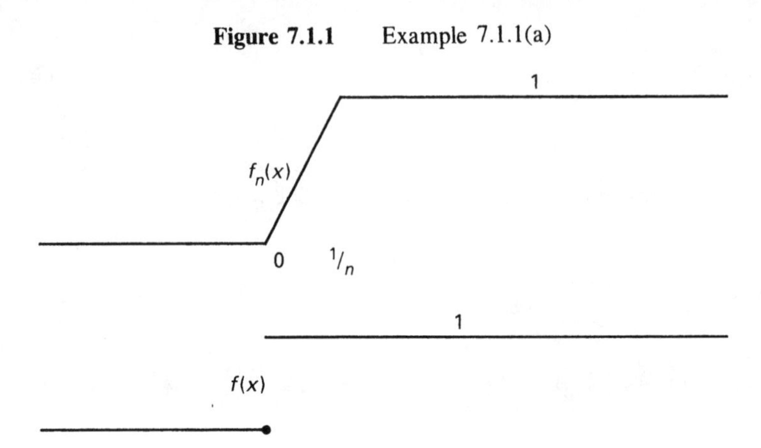

We obtain

$$\lim_{n\to\infty}\lim_{x\to 0} f_n(x) = \lim_{n\to\infty} 0 = 0.$$

But

$$\lim_{x\to 0}\lim_{n\to\infty} f_n(x) = \lim_{x\to 0} f(x),$$

which does not exist, since $f(x) \to 1$ as $x \to 0^+$, and $f(x) \to 0$ as $x \to 0^-$. Note also that each f_n is continuous everywhere, and f_n converges to f *pointwise*; that is, $f_n(x) \to f(x)$ for each x, but f is discontinuous at $x = 0$.

(b) Let

$$f_n(x) = \begin{cases} ne^{-nx}, & x > 0, \\ 0, & x \le 0. \end{cases}$$

In this case, $f_n \to f$ pointwise, where $f(x) = 0$ for all x. We consider the operations of integration and taking the limit as $n \to \infty$. We obtain

$$\lim_{n\to\infty}\int_0^\infty f_n(x)\,dx = \lim_{n\to\infty} 1 = 1,$$

but

$$\int_0^\infty \lim_{n\to\infty} f_n(x)\,dx = \int_0^\infty f(x)\,dx = 0.$$

Thus, although $f_n \to f$ pointwise, $\int_0^\infty f_n(x)\,dx \not\to \int_0^\infty f(x)\,dx$; in other words, the limit of the integral of f_n is not the integral of the limit of f_n.

(c) Let $f_n(x) = (1/n)\sin nx$; again $f_n \to f$ pointwise, where $f \equiv 0$. In this case we consider differentiation and the limit as $n \to \infty$. We have

$$\lim_{n\to\infty} \frac{d}{dx} f_n(x) = \lim_{n\to\infty} \cos nx,$$

which does not exist unless x is an integer multiple of 2π. But

$$\frac{d}{dx} \lim_{n\to\infty} f_n(x) = \frac{d}{dx} f(x) = \frac{d}{dx} 0 = 0.$$

Thus, $f_n \to f$ pointwise, but $df_n/dx \not\to df/dx$; that is, the limit of the derivative of f_n is not the derivative of the limit.

Pointwise convergence of f_n to f may be visualized geometrically via a *vertical line test*. For a fixed x, the vertical line through $(x, 0)$ travels a distance $|f_n(x) - f(x)|$ between the graphs of f_n and f at x. Equivalently, we can sketch $|f_n - f|$; the vertical line will then travel between 0 and $|f_n(x) - f(x)|$. This distance must approach 0 as $n \to \infty$. See Fig. 7.1.2 for the case $f_n(x) = ne^{-nx}, x > 0, f(x) \equiv 0$.

The idea of uniform convergence may be illustrated by a *horizontal line test*. For each $\epsilon > 0$, we draw the horizontal line through $(0, \epsilon)$ and ask whether the entire graph of $|f_n - f|$ will lie below the line for all sufficiently large n. If this is possible for all $\epsilon > 0$, then $f_n \to f$ uniformly. In the example illustrated in Fig. 7.1.2 we have $f_n(x) \to n$ as $x \to 0$, so there is no way to squash the entire graph of $|f_n - f|$ below ϵ; see Fig. 7.1.3.

7.1.2 Definitions

Let $f, f_1, f_2, \ldots$ be real–valued functions defined on the arbitrary set E. We say that f_n converges to f *pointwise* on E if for each $x \in E$, $f_n(x) \to f(x)$ as $n \to \infty$. We say that f_n converges to f *uniformly* on E if

$$\sup_{x\in E} |f_n(x) - f(x)| \to 0 \quad \text{as} \quad n \to \infty.$$

Figure 7.1.2 Vertical Line Test for Pointwise Convergence

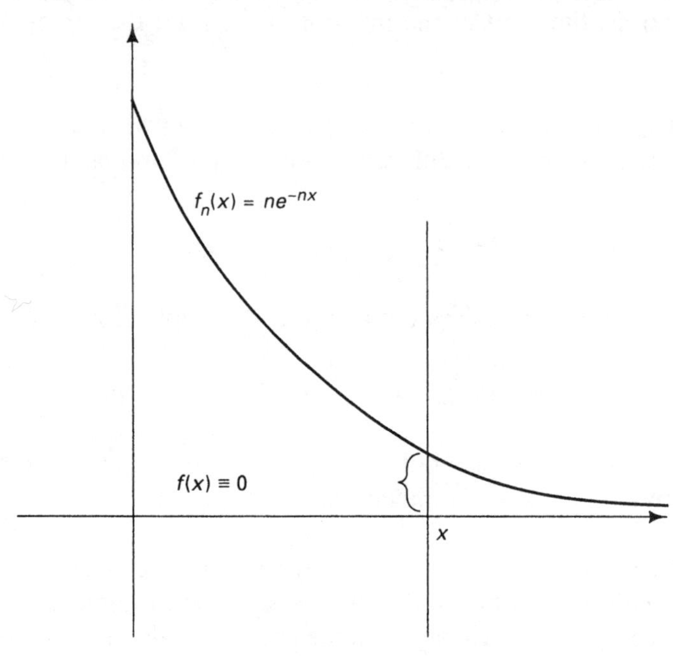

In other words, given $\epsilon > 0$ there is a positive integer N (depending only on ϵ) such that for all $x \in E$ and all $n \geq N$, $|f_n(x) - f(x)| < \epsilon$. Geometrically, the entire graph of $|f_n - f|$ is squashed below ϵ for all sufficiently large n.

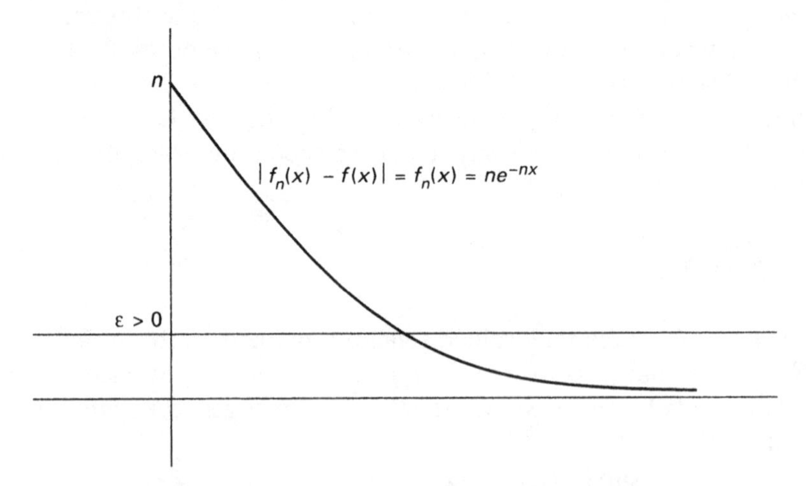

Figure 7.1.3 Horizontal Line Test for Uniform Convergence

7.1.3 Example

Let $f_n(x) = n^2 x(1-x)^n$, $0 \le x \le 1$; f_n converges pointwise to the zero function f. To check for uniform convergence, we will compute $\sup_{0\le x\le 1} |f_n(x) - f(x)|$; note that the sup is actually a maximum by continuity of $|f_n - f|$. If we take the derivative of $n^2x(1-x)^n$ and set the result equal to 0, we obtain $x = 1/(n+1)$; hence,

$$\sup_{0\le x\le 1} |f_n(x) - f(x)| = \frac{n^2}{n+1}(1 - \frac{1}{n+1})^n$$

$$= \frac{n^2}{n+1}\left[\left(1 - \frac{1}{n+1}\right)^{n+1}\right]^{n/n+1}.$$

Since $n/n+1 \to 1$, $n^2/n+1 \to \infty$, and $(1+r/m)^m \to e^r$ as $m \to \infty$ for any r, the above expression approaches infinity as $n \to \infty$. Thus, f_n does not converge uniformly to f.

We have already noted (see Fig. 7.1.2) that in Example 7.1.1(b), f_n does not converge uniformly to f. A similar analysis using a horizontal line test shows that the conclusion is the same in Example 7.1.1(a). In Example 7.1.1(c), f_n does converge uniformly to f (since $|f_n(x)| \le 1/n \to 0$), but this is not enough to ensure that $df_n/dx \to df/dx$.

Problems for Section 7.1

1. In Example 7.1.1(a), cut the domain of the f_n down to $(0, \infty)$, and evaluate $\lim_{x\to 0} \lim_{n\to\infty} f_n(x)$ and $\lim_{n\to\infty} \lim_{x\to 0} f_n(x)$.
2. Let $f_n(x) = x^n/(2+3x^n)$, $0 \le x \le 1$. Show that f_n converges pointwise, and determine whether the sequence converges uniformly on $[0, 1]$.
3. Let $f_n(x) = x^n/(n+x^n)$, $0 \le x \le 1$. Show that f_n converges pointwise, and determine whether or not the sequence converges uniformly on [0,1].
4. Show that the sequence of Problem 3 converges uniformly on $[1+\delta, \infty)$ for any $\delta > 0$, but does not converge uniformly on $[1, \infty)$.
5. The sequence $f_n(x) = x^n e^{-nx}$ converges pointwise to 0 on $[0, \infty)$. Does the sequence converge uniformly? Explain.

7.2 UNIFORM CONVERGENCE AND LIMIT OPERATIONS

We now discuss the relation between uniform convergence and interchange of limit operations.

7.2.1 THEOREM. *A uniform limit of continuous functions is continuous. That is, if E is a subset of a metric space, $f_1, f_2, \ldots$ are continuous real-valued functions on E, and $f_n \to f$ uniformly on E, then f is continuous on E. In particular, if $x_0 \in E$,*

$$\lim_{n\to\infty}\lim_{x\to x_0} f_n(x) = \lim_{x\to x_0}\lim_{n\to\infty} f_n(x) = f(x_0).$$

Proof. If $x_0 \in E$, then

$$|f(x) - f(x_0)| \le |f(x) - f_n(x)| + |f_n(x) - f_n(x_0)| + |f_n(x_0) - f(x_0)|.$$

Given $\epsilon > 0$, the uniform convergence of f_n to f allows us to find a positive integer N such that $|f(x) - f_n(x)| < \epsilon/3$ for all $n \ge N$ and all $x \in E$; set $x = x_0$ to obtain $|f_n(x_0) - f(x_0)| < \epsilon/3$. Fix $n \ge N$; since f_n is continuous at x_0, there is a $\delta > 0$ such that $|f_n(x) - f_n(x_0)| < \epsilon/3$ whenever $d(x, x_0) < \delta$. Thus, $d(x, x_0) < \delta$ implies $|f(x) - f_n(x)| < \epsilon$. ■

The next two results relate uniform convergence to interchange of the limit operation with integration and differentiation. The results can be proved in somewhat greater generality, but we have strengthened the hypotheses in order to simplify the proofs.

7.2.2 THEOREM. *Let $f_1, f_2, \ldots$ be continuous real-valued functions on $[a, b]$, and let α be of bounded variation on $[a, b]$. If $f_n \to f$ uniformly on $[a, b]$, then*

$$\int_a^b f_n\, d\alpha \to \int_a^b f\, d\alpha \quad \text{as} \quad n \to \infty.$$

Proof. By Theorem 7.2.1, f is continuous on $[a, b]$, so $\int_a^b f\, d\alpha$ exists. Now

$$\begin{aligned}
\left| \int_a^b f_n\, d\alpha - \int_a^b f\, d\alpha \right| &= \left| \int_a^b (f_n - f)\, d\alpha \right| \\
&\leq \sup_{a \leq x \leq b} |f_n(x) - f(x)| V(\alpha) \\
&\qquad \text{by a standard bound (see Section 6.4.4)} \\
&\to 0 \quad \text{as} \quad n \to \infty,
\end{aligned}$$

since $f_n \to f$ uniformly. ■

7.2.3 THEOREM. *Assume that for each $n = 1, 2 \ldots,$ f_n has a continuous derivative f_n' on $[a, b]$, and that for some $x_0 \in [a, b]$, $f_n(x_0)$ converges to a finite limit. If f_n' converges uniformly on $[a, b]$, then f_n converges uniformly on $[a, b]$ to a limit function f, and $f_n'(x) \to f'(x)$ for every $x \in [a, b]$.*

The purpose of the hypothesis that $f_n(x_0)$ converges to a finite limit is to exclude cases such as $f_n(x) = c_n$ for all x, where f_n', which is identically 0, converges uniformly, but f_n need not converge uniformly.

Proof. Suppose $f_n' \to g$ uniformly on $[a, b]$. Then for $a \leq x \leq b$ we have

$$\begin{aligned}
(1) \qquad f_n(x) - f_n(x_0) &= \int_{x_0}^x f_n'(t)\, dt \qquad \text{by Theorem 6.2.4(b)} \\
&\to \int_{x_0}^x g(t)\, dt \qquad \text{by Theorem 7.2.2.}
\end{aligned}$$

Since $f_n(x_0)$ converges by hypothesis and $f_n(x) - f_n(x_0)$ converges by the above observation, $f_n(x)$ converges; call the limit $f(x)$. Let $n \to \infty$ in (1) to obtain

$$(2) \qquad f(x) - f(x_0) = \int_{x_0}^x g(t)\, dt.$$

By Theorem 6.2.4(a), $f'(x) = g(x)$ on $[a, b]$, so $f_n' \to f'$ uniformly on $[a, b]$. To establish uniform convergence of f_n, note that

$$\begin{aligned}
|f_n(x) - f(x)| &\le |f_n(x) - f_n(x_0) - (f(x) - f(x_0))| \\
&\quad + |f_n(x_0) - f(x_0)| \\
&= \left| \int_{x_0}^{x} [f_n'(t) - f'(t)]\, dt \right| + |f_n(x_0) - f(x_0)| \\
&\quad \text{by (1) and (2)} \\
&\le \sup_{a \le t \le b} |f_n'(t) - f'(t)|(b - a) + |f_n(x_0) - f(x_0)| \\
&\quad \text{(Section 6.4.4)} \\
&\to 0 \quad \text{as} \quad n \to \infty, \text{ uniformly in } \ x.
\end{aligned}$$ ■

The following result is occasionally useful in establishing uniform convergence.

7.2.4 LEMMA. *Suppose the real-valued functions $f_1, f_2, \ldots$ have the uniform Cauchy property on E; that is, given $\epsilon > 0$ there is a positive integer N (depending only on ϵ) such that for $n, m \ge N$ we have*

$$|f_n(x) - f_n(x)| < \epsilon \qquad \textit{for every} \quad x \in E.$$

Then the sequence $\{f_n\}$ converges uniformly on E.

Proof. For each $x \in E$, $\{f_n(x)\}$ is a Cauchy sequence of real numbers; hence, $f_n(x)$ converges to a limit $f(x)$. If $|f_n(x) - f_n(x)| < \epsilon$ for all n, $m \ge N$ and all $x \in E$, fix n and let $m \to \infty$ to conclude that $|f_n(x) - f(x)| \le \epsilon$ for all $n \ge N$ and all $x \in E$. Thus, $f_n \to f$ uniformly on E. ■

We have seen that a uniform limit of continuous function is continuous. Under certain conditions, pointwise convergence of a sequence of continuous functions to a continuous limit implies uniform convergence.

7.2.5 Dini's Theorem

Let $f_1, f_2, \ldots$ be continuous real-valued functions on the compact set E, and assume the f_n form a monotone sequence (either $f_{n+1}(x) \le f_n(x)$

for all $x \in E$ *and all* $n = 1, 2, \ldots$, *or* $f_{n+1}(x) \geq f_n(x)$ *for all* $x \in E$ *and all* $n = 1, 2, \ldots$). *If* $f_n \to f$ *pointwise on* E *and* f *is continuous on* E, *then* $f_n \to f$ *uniformly on* E.

Proof. We may assume that the f_n form a decreasing sequence (in the increasing case, simply consider the functions $-f_n$, which decrease). If $g_n = f_n - f$, the g_n form a decreasing sequence of nonnegative continuous functions converging pointwise to 0. If $\epsilon > 0$, let $V_n = \{x \in E : g_n(x) < \epsilon\}$, an open set by continuity of g_n. If $x \in E$, then $g_n(x) < \epsilon$ eventually; hence, $\bigcup_{n=1}^{\infty} V_n = E$. Since E is compact, $\bigcup_{n=1}^{N} V_n = E$ for some N. But since the g_n decrease, we have $V_n \subseteq V_{n+1}$ for all n, and therefore $\bigcup_{n=1}^{N} V_n = V_N$. Thus, if $x \in E$, then $x \in V_N$; that is, $g_N(x) < \epsilon$. If $n \geq N$, we have $0 \leq g_n(x) \leq g_N(x) < \epsilon$. Thus, $g_n \to 0$ uniformly on E. ■

Problems for Section 7.2

1. In Theorem 7.2.3, drop the hypothesis that for some $x_0 \in [a, b]$, $f_n(x_0)$ converges to a finite limit. Show that the result no longer holds.
2. If $\{f_n\}$ and $\{g_n\}$ converge uniformly on E, show that $\{f_n + g_n\}$ converges uniformly on E.
3. Give an example in which $\{f_n\}$ and $\{g_n\}$ converge uniformly on E, but $\{f_n g_n\}$ fails to converge uniformly on E.
4. If $\{f_n\}$ converges uniformly on E and each f_n is bounded (i.e., for some $M_n > 0$, $|f_n(x)| \leq M_n$ for every $x \in E$), show that the f_n are *uniformly bounded* on E; that is, for some $M > 0$, $|f_n(x)| \leq M$ for all $x \in E$ and all $n = 1, 2, \ldots$.
5. In Problem 4, drop the assumption that each f_n is bounded, and show that the result no longer holds.

7.3 THE WEIERSTRASS M-TEST AND APPLICATIONS

Uniform convergence of a series of functions $\sum_{n=1}^{\infty} f_n$ means, by definition, uniform convergence of the sequence of nth partial sums $s_n = \sum_{k=1}^{n} f_k$. Most of the time it is difficult to evaluate the sum of an infinite series, so it is useful to have sufficient conditions for uniform convergence that involve only the functions f_n. The following result is widely used.

7.3.1 Weierstrass M–Test

Let $f_1, f_2, \ldots$ be real–valued functions on the set E. If $|f_n(x)| \leq M_n$ for all $x \in E$ and all $n = 1, 2, \ldots$, where $\sum_{n=1}^{\infty} M_n < \infty$, then the series $\sum_{n=1}^{\infty} f_n$ converges uniformly on E. Thus, if each f_n is continuous on E, then (by Theorem 7.2.1) $\sum_{n=1}^{\infty} f_n$ is continuous on E.

Proof. If s_n is the nth partial sum of the series, then for $m < n$,

$$|s_n(x) - s_m(x)| = \left| \sum_{k=m+1}^{n} f_k(x) \right| \leq \sum_{k=m+1}^{n} |f_k(x)| \leq \sum_{k=m+1}^{n} M_k \to 0 \quad \text{as} \quad m, n \to \infty,$$

since $\sum_{n=1}^{\infty} M_n < \infty$. Thus, the sequence $\{s_n\}$ has the uniform Cauchy property and therefore converges uniformly by Lemma 7.2.4. ■

7.3.2 Example

Let $f_n(x) = (\sin nx)/n^2$, $x \in R$. Then $|f_n(x)| \leq M_n = 1/n^2$ for all x, and it follows that $\sum_{n=1}^{\infty} \sin nx/n^2$ converges uniformly on R. (Unfortunately, the Weierstrass M–test does not help us find the limit function explicitly.)

We now use the Weierstrass M–test to produce a rather spectacular application of uniform convergence.

7.3.3 An Everywhere Continuous, Nowhere Differentiable Function

Let $f(x) = |x|$, $-1 \leq x \leq 1$, and extend f periodically to all of R. In other words, the extension is required to satisfy $f(x + 2) = f(x)$ for

Figure 7.3.1 $f(x) = |x|$, Extended Periodically

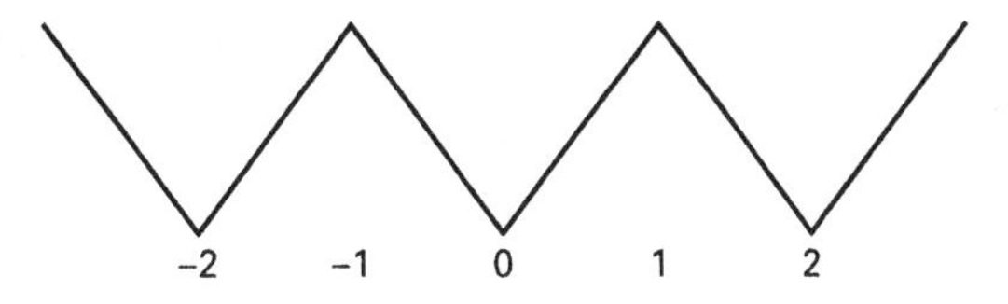

all x. The resulting "sawtooth" curve is sketched in Fig. 7.3.1. Now if x and y both belong to the same branch (linear segment) of the curve, then $|f(y) - f(x)| = |y - x|$. In general, as shown in Fig. 7.3.2,

$$|f(y) - f(x)| \leq |y - x|. \tag{1}$$

Fix the point $x \in R$ and the positive integer m. Let $h_m = \pm\frac{1}{2}4^{-m}$, where the choice of + or − will be decided in a moment. Since $|4^m(x + h_m) - 4^m x| = 4^m|h_m| = \frac{1}{2}$, either there will be no integer between $4^m(x - \frac{1}{2}4^{-m})$ and $4^m x$ or there will be no integer between $4^m x$ and $4^m(x + \frac{1}{2}4^{-m})$; see Fig. 7.3.3. We choose + or − so that no integer will lie between $4^m x$ and $4^m(x + h_m)$; in Fig. 7.3.3 we would choose +. Let

$$c_n = \frac{f[4^n(x + h_m)] - f(4^n x)}{h_m}, \qquad n = 0, 1, 2, \ldots.$$

We claim that

$$c_n = 0, \quad n > m; \qquad |c_m| = 4^m; \qquad |c_n| \leq 4^n, \quad n < m. \tag{2}$$

For if $n > m$, $|4^n h_m| = \frac{1}{2}4^{n-m}$, an even integer. Since f has period 2, $c_n = 0$. If $n < m$, (1) gives $|c_n| \leq 4^n|h_m|/|h_m| = 4^n$. If $n = m$,

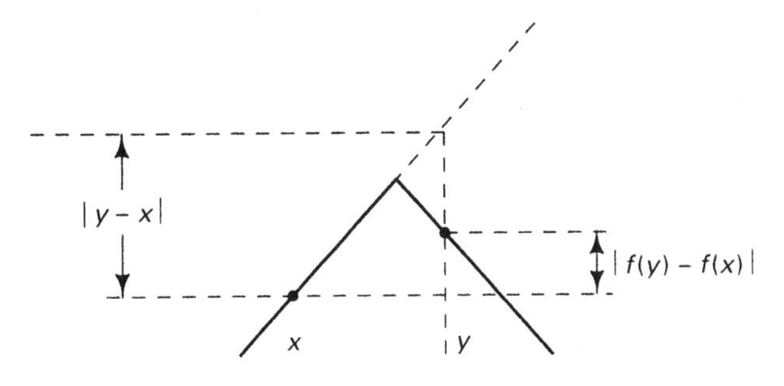

Figure 7.3.2 An Inequality for the Sawtooth Function

Figure 7.3.3 Determining the Choice of h_m

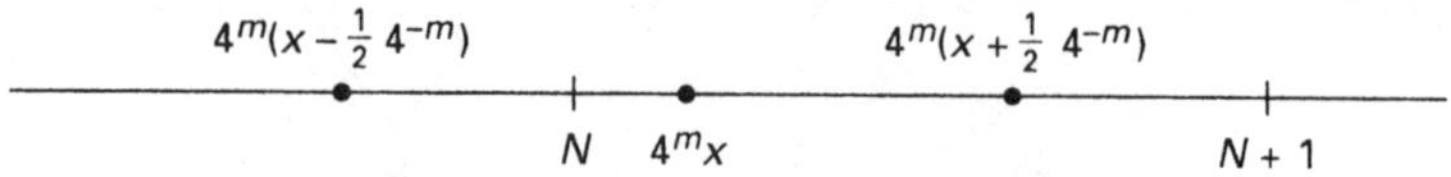

$4^m(x - \frac{1}{2}4^{-m})$ $\quad$ $4^m(x + \frac{1}{2}4^{-m})$

$N \quad 4^m x \quad N+1$

the fact that no integer lies between $4^m x$ and $4^m(x + h_m)$ means that these two points belong to the same branch of f, so $|c_m| = |4^m(x + h_m) - 4^m x|/|h_m| = 4^m$, as desired.

We are now ready for the construction of the desired function. Define

$$g(x) = \sum_{n=0}^{\infty} \left(\frac{3}{4}\right)^n f(4^n x), \qquad x \in R.$$

(Intuitively, as more terms are added to the sum, more "spikes" are created; see Fig. 7.3.4.) Since $|f| \leq 1$ and $\sum_{n=0}^{\infty} \left(\frac{3}{4}\right)^n < \infty$, g is continuous on R by the Weierstrass M-test and Theorem 7.2.1. If x is any real number, we show that g is not differentiable at x. If m is any positive integer,

$$\begin{aligned} \left|\frac{g(x+h_m) - g(x)}{h_m}\right| &= \left|\sum_{n=0}^{\infty} \left(\frac{3}{4}\right)^n c_n\right| \\ &= \left|\left(\frac{3}{4}\right)^m c_m + \sum_{n=0}^{m-1} \left(\frac{3}{4}\right)^n c_n\right| \qquad \text{by (2).} \end{aligned}$$

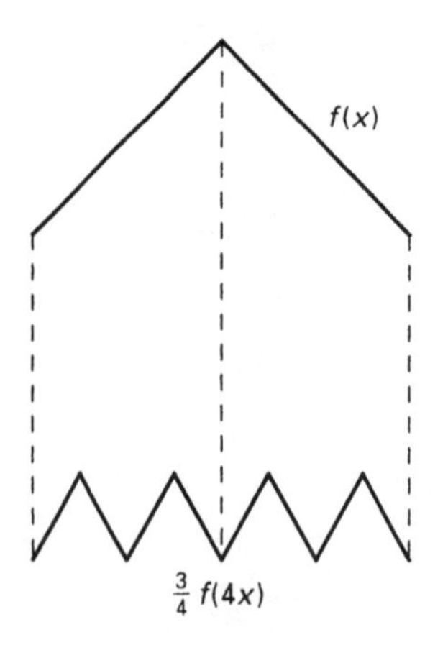

Figure 7.3.4 Constructing an Everywhere Continuous, Nowhere Differentiable Function

Now

$$\left| \left(\frac{3}{4}\right)^m c_m \right| = 3^m$$

and

$$\left| \sum_{n=0}^{m-1} \left(\frac{3}{4}\right)^n c_n \right| \leq \sum_{n=0}^{m-1} 3^n = \frac{1}{2}(3^m - 1) < 3^m;$$

hence,

$$\begin{aligned} \left| \frac{g(x+h_m) - g(x)}{h_m} \right| &\geq 3^m - \frac{1}{2}(3^m - 1) \\ &= \frac{1}{2}(3^m + 1) \\ &\to \infty \quad \text{as} \quad m \to \infty. \end{aligned}$$

Since $h_m \to 0$ as $m \to \infty$, it follows that g is not differentiable at x.

Problems for Section 7.3

1. Show that the series $\sum_{n=0}^{\infty} e^{-nx}$ converges uniformly on $[a, \infty)$ for any $a > 0$.
2. Show that $\sum_{n=0}^{\infty} e^{-nx}$ does not converge uniformly on $(0, \infty)$.
3. Suppose the power series $\sum_{n=0}^{\infty} a_n x^n$ converges at $x = r$. If $|a| < |r|$, show that the series converges absolutely at $x = a$.
4. Continuing Problem 3, show that if $0 < a < |r|$, then the series converges uniformly on $[-a, a]$.
5. Continuing Problem 4, show that if the power series $\sum_{n=0}^{\infty} a_n x^n$ has radius of convergence r, the series may be integrated and differentiated term by term (as if it were an ordinary polynomial) for $-r < x < r$.
6. If $f_1, f_2, \ldots$ are continuous real-valued functions on $[a, b]$ and $\sum_{n=1}^{\infty} f_n$ converges uniformly on $[a, b]$, show that for any α of bounded variation on $[a, b]$, we have

$$\int_a^b \left(\sum_{n=1}^{\infty} f_n \right) d\alpha = \sum_{n=1}^{\infty} \int_a^b f_n \, d\alpha.$$

7.4 EQUICONTINUITY AND THE ARZELA–ASCOLI THEOREM

We know that any bounded sequence of real numbers has a subsequence converging to a limit in R. A similar question will now be raised for sequences of functions. When will there exist a uniformly convergent subsequence? First, we prove some preliminary results.

7.4.1 LEMMA. *Let A be a subset of a metric space with the property that every open covering of A has a countable subcovering. (Every subset of R^p has this property; see the proof of Theorem 2.2.5.) Then A has a countable dense subset; in other words, there is a countable set $E \subseteq A$ such that each $x \in A$ can be expressed as a limit of a sequence of points in E. Conversely, if A has a countable dense subset, then every open covering of A has a countable subcovering.*

Proof. For each $n = 1, 2, \ldots$, we have $A \subseteq \bigcup_{x \in A} B_{1/n}(x)$; hence, A is covered by a countable union of balls of radius $1/n$ and centers in A, say $A \subseteq \bigcup_{j=1}^{\infty} B_{1/n}(x_{nj})$. Let $E = \{x_{nj} : n, j = 1, 2, \ldots\}$. If $\epsilon > 0$ and $x \in A$, choose n so large that $1/n < \epsilon$. Since $x \in B_{1/n}(x_{nj})$ for some j, we have $d(x, x_{nj}) < \epsilon$. It follows that E is dense.

If A has a countable dense subset E, we may reproduce the first part of the proof of the Heine–Borel Theorem 2.2.5, using E instead of the rationals in R^p, to show that every open covering of A has a countable subcovering. ■

If A is an interval of reals, we may take E to consist of the rational numbers in A. But this idea does not work for an arbitrary set $A \subseteq R$. If, for example, A is the set of irrationals, the set of rationals of A is empty, so is certainly not dense. However, the procedure given in the proof of Lemma 7.4.1 can be used for an arbitrary $A \subseteq R^p$.

7.4.2 LEMMA. *Let $\{f_n\}$ be a sequence of real–valued functions on the countable set $E = \{x_1, x_2, \ldots\}$, and assume that for each $j = 1, 2, \ldots$, $\{f_n(x_j) : n = 1, 2, \ldots\}$ is bounded. Then $\{f_n\}$ has a subsequence that converges (to a finite limit) at each point of E.*

Proof. The sequence $f_1(x_1), f_2(x_1), f_3(x_1), \ldots$ is bounded; hence, by Theorem 2.3.2 there is a subsequence, call it $\{f_{11}, f_{12}, f_{13}, \ldots\}$, of $\{f_n\}$ such that $f_{1n}(x_1)$ converges as $n \to \infty$. Now $\{f_{11}(x_2),$

$f_{12}(x_2), f_{13}(x_2), \dots\}$ is bounded, so there is a subsequence $f_{21}, f_{22}, f_{23}, \dots$ of $\{f_{1n}\}$ such that $f_{2n}(x_2)$ converges as $n \to \infty$. Continue inductively in this fashion; the results may be summarized by the following array:

$$
\begin{array}{cccc}
f_{11} & f_{12} & f_{13} & \cdots \\
f_{21} & f_{22} & f_{23} & \cdots \\
 & \vdots & & \\
f_{n1} & f_{n2} & f_{n3} & \cdots \\
 & \vdots & &
\end{array}
$$

We form the *diagonal subsequence* $g_n = f_{nn}$, $n = 1, 2, \dots$. To check that we have a legal subsequence, note that row $k + 1$ of the above array is a subsequence of row k, and therefore $f_{k+1,k+1}$ appears in row k somewhere to the right of f_{kk}. Similarly, $f_{k+2,k+2}$ appears in row k to the right of $f_{k+1,k+1}$. Thus, $\{g_n, n \geq k\}$ is a subsequence of $\{f_{kj}, j = 1, 2, \dots\}$; in particular, $\{g_n, n \geq 1\}$ is a subsequence of $\{f_n\}$. Furthermore, since $f_{kj}(x_k)$ converges as $j \to \infty$, $g_n(x_k)$ converges as $n \to \infty$ $(k = 1, 2, \dots)$. Thus, $\{g_n\}$ converges pointwise on E. ■

Recall that the function $f: \Omega \to \Omega'$ is uniformly continuous on Ω if for every $\epsilon > 0$ there is a $\delta > 0$ such that whenever $x, y \in \Omega$ and $d(x, y) < \delta$, we have $d(f(x), f(y)) < \epsilon$; δ depends only on ϵ but not on the particular x and y. We now consider a generalization of this concept.

7.4.3 Definition

Let $\{f_i\}$ be an arbitrary (possibly uncountable) family of functions from the metric space Ω to the metric space Ω'. If $E \subseteq \Omega$, $\{f_i\}$ is said to be *equicontinuous on* E if for every $\epsilon > 0$ there is a $\delta > 0$ such that for all $x, y \in E$ and all i,

$$d(x, y) < \delta \quad \text{implies} \quad d(f_i(x), f_i(y)) < \epsilon.$$

Thus, not only is δ independent of the points x and y, but δ does not depend on the particular f_i in the family either.

The central result about equicontinuity is the following.

7.4.4 Arzela–Ascoli Theorem

Let $\{f_n\}$ be a sequence of real-valued functions that is equicontinuous on the compact set K. If the sequence is pointwise bounded (i.e., $\sup_n |f_n(x)| < \infty$ for each $x \in K$), then there is a subsequence converging uniformly on K. Furthermore, the sequence $\{f_n\}$ is uniformly bounded; in other words, $\sup_{n,x} |f_n(x)| < \infty$.

Proof. By Lemma 7.4.1, K has a countable dense subset E. Since $\{f_n\}$ is pointwise bounded, the hypothesis of Lemma 7.4.2 is satisfied; hence, there is a subsequence $\{g_n\}$ converging pointwise on E; we must show that $\{g_n\}$ converges uniformly on K. Given $\epsilon > 0$, choose $\delta > 0$ so that if $x, y \in K$ and $d(x,y) < \delta$ then $d(f_n(x), f_n(y)) < \epsilon/3$ for all n. If $y \in K$, then since E is dense there is a point $x \in E$ such that $d(x,y) < \delta$; that is, $y \in B_\delta(x)$. It follows that $K \subseteq \bigcup_{x\in E} B_\delta(x)$, and therefore, by compactness, $K \subseteq \bigcup_{i=1}^{p} B_\delta(x_i)$ for some $x_1, \ldots, x_p \in E$. Since $g_n(x_i)$ converges as $n \to \infty$ for each i, there is a positive integer N such that $|g_n(x_i) - g_m(x_i)| < \epsilon/3$ for all $n, m \geq N$ and all $i = 1, \ldots, p$. If $x \in K$, then $x \in B_\delta(x_i)$ for some i; thus,

$$|g_n(x) - g_m(x)| \leq |g_n(x) - g_n(x_i)| + |g_n(x_i) - g_m(x_i)| + |g_m(x_i) - g_m(x)|.$$

The first and third terms on the right side are less than $\epsilon/3$ by equicontinuity, and the second term is less than $\epsilon/3$ provided $n, m \geq N$. It follows that $|g_n(x) - g_m(x)| < \epsilon$ for all $n, m \geq N$ and all $x \in K$. By Lemma 7.2.4, $\{g_n\}$ converges uniformly on K.

To prove uniform boundedness, let $h(x) = \sup_n |f_n(x)|$. If $x, y \in K$ and $d(x,y) < \delta$, we have, for each $n = 1, 2, \ldots$,

$$|f_n(y)| \leq |f_n(y) - f_n(x)| + |f_n(x)| < \epsilon + |f_n(x)|.$$

Take the sup over n to obtain $h(y) \leq h(x) + \epsilon$; by symmetry, $h(x) \leq h(y) + \epsilon$; hence, $|h(x) - h(y)| < \epsilon$. Thus, h is a continuous function on the compact set K, so h is bounded; in other words, $\sup_{n,x} |f_n(x)| < \infty$, as desired. ■

If the functions f_n are not real–valued but take values in an arbitrary metric space, the conclusion of Theorem 7.4.4 still holds if the pointwise boundedness hypothesis is replaced by the assumption that for each $x \in K$, $\{f_n(x) : n = 1, 2, \dots\}$ is a subset of a compact set, and hence the sequence $f_1(x), f_2(x), \dots$ has a convergent subsequence. This allows the construction given in the proof of Lemma 7.4.2 to be retained, and the proof of Theorem 7.4.4 to be essentially reproduced.

Problems for Section 7.4

1. Give an example of a sequence of real-valued functions f_n on a countable set $E = \{x_1, x_2, \dots\}$ such that $\{f_n\}$ is pointwise bounded but not uniformly bounded.

2. Define, for each $n = 1, 2, \dots,$

$$f_n(x) = 2^n \left(x - \frac{k}{2^n}\right), \qquad \frac{k}{2^n} < x \le \frac{k+1}{2^n}, \qquad k = 0, 1, ..., 2^n - 1.$$

Take $f_n(0) = 0$ for all n.

(a) Sketch $f_n(x)$, $0 \le x \le 1$, for $n = 1, 2$.

(b) Show that $\{f_n\}$ is uniformly bounded and converges pointwise on a countable dense subset of $[0, 1]$, but does not converge pointwise on $[0, 1]$.

3. Let $\{f_n\}$ be an equicontinuous sequence of functions on the compact set K. If $f_n \to f$ pointwise on K, show that $f_n \to f$ uniformly on K.

4. Let $f_1, f_2, \dots$. be continuous functions on the compact set K, and assume $f_n \to f$ uniformly on K.

(a) Show that $\{f_n\}$ is equicontinuous on K.

(b) If $x_1, x_2, \dots \in K$ and $x_n \to x$, show that $f_n(x_n) \to f(x)$.

(c) Now remove the hypothesis that K is compact. Show that (a) no longer holds, but (b) is still valid.

7.5 THE WEIERSTRASS APPROXIMATION THEOREM

We are going to investigate the problem of approximating a continuous function by polynomials. The main result, the Weierstrass Approximation Theorem, is very well motivated by considerations of basic probability theory. As we go along, we will make parenthetical comments that should be helpful for those who have some familiarity with probability. The proofs, however, do not use probability considerations. We need the following two results.

7.5.1 LEMMA. *If* $0 \leq x \leq 1$,

$$\sum_{k=0}^{n} \left(x - \frac{k}{n}\right)^2 \binom{n}{k} x^k (1-x)^{n-k} = \frac{x(1-x)}{n},$$

where

$$\binom{n}{k} = \frac{n!}{k!(n-k)!}.$$

(If X is the number of heads in n independent tosses of a coin with probability of heads x on a given toss, E stands for expectation or average value, and Var indicates variance, the above expression is

$$\begin{aligned} E\left[\left(\frac{X}{n} - x\right)^2\right] &= \frac{1}{n^2} E[(X - nx)^2] \\ &= \frac{1}{n^2} \operatorname{Var} X \\ &= \frac{nx(1-x)}{n^2} \\ &= \frac{x(1-x)}{n}. \end{aligned}$$
)

Proof. By the binomial theorem,

$$\sum_{k=0}^{n} \binom{n}{k} x^k (1-x)^{n-k} = (x + 1 - x)^n = 1.$$

Also,

$$\begin{aligned}\sum_{k=0}^{n} k\binom{n}{k}x^k(1-x)^{n-k} &= \sum_{k=1}^{n} \frac{kn!}{k!(n-k)!}xx^{k-1}(1-x)^{n-k}\\ &= nx\sum_{k=1}^{n}\frac{(n-1)!}{(k-1)!(n-k)!}x^{k-1}(1-x)^{n-1-(k-1)}\\ &= nx.\end{aligned}$$

Finally, since $k^2 = k(k-1)+k$,

$$\begin{aligned}\sum_{k=0}^{n} k^2\binom{n}{k}x^k(1-x)^{n-k} &= nx + \sum_{k=0}^{n} k(k-1)\binom{n}{k}x^k(1-x)^{n-k}\\ &= nx + n(n-1)x^2\sum_{k=2}^{n}\frac{(n-2)!}{(k-2)!(n-k)!}x^{k-2}(1-x)^{n-2-(k-2)}\\ &= nx + n(n-1)x^2 = nx(1-x)+n^2x^2.\end{aligned}$$

Thus,

$$\begin{aligned}\sum_{k=0}^{n}(x-\frac{k}{n})^2\binom{n}{k}x^k(1-x)^{n-k} &= x^2 - \frac{2x}{n}(nx) + \frac{1}{n^2}(nx(1-x)+n^2x^2) = \frac{x(1-x)}{n}. \quad \blacksquare\end{aligned}$$

7.5.2 LEMMA. *Let $x \in [0,1]$, and let n be a positive integer. If S is the set of integers $k \in \{0,1,\ldots,n\}$ such that $|x-k/n| \geq n^{-1/4}$, then*

$$\sum_{k\in S}\binom{n}{k}x^k(1-x)^{n-k} \leq \frac{1}{4\sqrt{n}}.$$

(If X is the random variable given in Lemma 7.5.1, then the probability that $|X/n - x| \le n^{-1/4}$ is, by Chebyshev's inequality, $\le E[(X/n - x)^2]/n^{-1/2} = x(1-x)/nn^{-1/2}$, which is at most $1/4\sqrt{n}$.)

Proof. The sum to be estimated may be written as

$$\sum_{k \in S} \frac{(x - k/n)^2}{(x - k/n)^2} \binom{n}{k} x^k (1-x)^{n-k}$$

$$\le \sqrt{n} \sum_{k \in S} \left(x - \frac{k}{n}\right)^2 \binom{n}{k} x^k (1-x)^{n-k}$$

$$\le \frac{x(1-x)}{\sqrt{n}} \qquad \text{by Lemma 7.5.1.}$$

But $x(1-x)$ is maximum at $x = \frac{1}{2}$, and the result follows. ■

We are now ready for the main result.

7.5.3 Weierstrass Approximation Theorem

Let f be a continuous real–valued function on $[0, 1]$. Define the Bernstein polynomials for f by

$$B_n(x) = \sum_{k=0}^{n} f\left(\frac{k}{n}\right) \binom{n}{k} x^k (1-x)^{n-k}, \qquad 0 \le x \le 1.$$

Then $B_n \to f$ uniformly on $[0, 1]$, so that f can be uniformly approximated by polynomials.

(If X is the random variable given in Lemma 7.5.1, then $B_n(x) = E[f(X/n)]$. For large n, X/n will (with high probability) be close to x by the Law of Large Numbers. Since f is continuous, $f(X/n)$ should be close to $f(x)$. This suggests convergence in some sense of B_n to f.)

Proof. By Theorem 4.2.4, f is uniformly continuous on $[0, 1]$. Thus, given $\epsilon > 0$, there is a $\delta > 0$ such that $|x - y| < \delta$ implies $|f(x) - f(y)| < \epsilon/2$. By Corollary 4.2.2, f is bounded, say $|f| \le M$. Now

$$f(x) - B_n(x) = \sum_{k=0}^{n} \left[f(x) - f\left(\frac{k}{n}\right)\right] \binom{n}{k} x^k (1-x)^{n-k}.$$

Thus,

$$|f(x) - B_n(x)| \leq \sum_{k=0}^{n} \left| f(x) - f\left(\frac{k}{n}\right) \right| \binom{n}{k} x^k (1-x)^{n-k}.$$

If we sum only over those k for which $|x - k/n| \geq n^{-1/4}$, the result is at most $2M/4\sqrt{n} = M/2\sqrt{n}$ by Lemma 7.5.2. This can be made less than $\epsilon/2$ provided $n > M^2/\epsilon^2$. If $|x - k/n| < n^{-1/4}$ and $n^{-1/4} < \delta$ (in other words, $n > \delta^{-4}$), then $|x - k/n| < \delta$, so $|f(x) - f(k/n)|$ will be less than $\epsilon/2$. Thus, if we sum over those k for which $|x - k/n| < n^{-1/4}$, the result is at most $\epsilon/2$. It follows that if $n > \max(M^2/\epsilon^2, \delta^{-4})$, we have $|f(x) - B_n(x)| < \epsilon$ for all $x \in [0,1]$. ■

There is no difficulty in approximating a continuous real–valued function on an arbitrary closed bounded interval $[a,b]$ by polynomials; see Problem 1.

The Weierstrass approximation theorem has been generalized to the following result, known as the *Stone–Weierstrass theorem*.

Let A be an algebra of continuous real–valued functions on the compact set K. ("Algebra" means that if $f, g \in A$ then $f + g \in A$ and $fg \in A$; also, if $f \in A$ and $c \in R$, then $cf \in A$.) *Assume that A contains all constant functions and separates points; that is, if $x, y \in K$, $x \neq y$, there is an $f \in A$ with $f(x) \neq f(y)$. Then for any continuous $f: K \to R$ and any $\epsilon > 0$ there is a $g \in A$ such that $|f(x) - g(x)| < \epsilon$ for all $x \in K$. Thus, f can be uniformly approximated by functions in A.* (In the case of the Weierstrass approximation theorem, $K = [0,1]$ and A is the set of all polynomials.)

Problems for Section 7.5

1. Show that a continuous real-valued function f on $[a,b]$ can be uniformly approximated by polynomials.

The following problems constitute a project involving the reversal of the order of summation of a double series.

2. Give an example of a double sequence of real numbers a_{nj}, $n, j = 1, 2, \ldots$, such that $\sum_{n=1}^{\infty} \sum_{j=1}^{\infty} a_{nj} \neq \sum_{j=1}^{\infty} \sum_{n=1}^{\infty} a_{nj}$.

3. Let $x_k = 1/k$, $k = 1, 2, \ldots$, $x_0 = 0$, and form the set $E = \{x_0, x_1, x_2, \ldots\}$. Let $\{a_{nj}\}$ be a double sequence of reals such that $\sum_{n=1}^{\infty} \sum_{j=1}^{\infty} |a_{nj}| < \infty$. (This is assumed in Problems 4, 5, 6, and 7 also.) Define

$$f_n(x_k) = \sum_{j=1}^{k} a_{nj}, \qquad k \geq 1,$$

$$f_n(x_0) = \sum_{j=1}^{\infty} a_{nj}.$$

Use the Weierstrass M-test to show that $\sum_{n=1}^{\infty} f_n$ converges uniformly on E.

4. Show that each f_n is continuous on E; hence, by Theorem 7.2.1, $f = \sum_{n=1}^{\infty} f_n$ is continuous on E.

5. Show that for all $k \geq 1$,

$$\sum_{n=1}^{\infty} \sum_{j=1}^{k} a_{nj} = \sum_{j=1}^{k} \sum_{n=1}^{\infty} a_{nj}.$$

6. Show that

$$\sum_{n=1}^{\infty} \sum_{j=1}^{\infty} a_{nj} = \lim_{k \to \infty} \sum_{n=1}^{\infty} \sum_{j=1}^{k} a_{nj}.$$

7. Finally, show that if $\sum_{n=1}^{\infty} \sum_{j=1}^{\infty} |a_{nj}| < \infty$, then

$$\sum_{n=1}^{\infty} \sum_{j=1}^{\infty} a_{nj} = \sum_{j=1}^{\infty} \sum_{n=1}^{\infty} a_{nj} \qquad \text{(finite)}.$$

(Note that the same result holds under the hypothesis that $\sum_{j=1}^{\infty} \sum_{n=1}^{\infty} |a_{nj}| < \infty$; to see this, simply set $b_{jn} = a_{nj}$.)

8. Let $\{a_{nj}\}$ be a nonnegative double sequence of real numbers. Define

$$s_{Nk} = \sum_{n=1}^{N} \sum_{j=1}^{k} a_{nj}, \qquad s = \sup\{s_{Nk} : N, k = 1, 2, \ldots\}.$$

Show that

$$\sum_{n=1}^{\infty}\sum_{j=1}^{\infty} a_{nj} = \sum_{j=1}^{\infty}\sum_{n=1}^{\infty} a_{nj} = s.$$

Thus, the order of summation in a nonnegative double series may always be reversed. If the series diverges to ∞ when summed in one order, it will diverge to ∞ in the other order as well.

(If s is finite, the result follows from Problem 7, but a general argument not based on Problem 7 may be given.)

9. Suppose $f(x) = \sum_{n=0}^{\infty} a_n x^n$ and $g(x) = \sum_{n=0}^{\infty} b_n x^n$ converge (at least) for $|x| < r$. Define the *Cauchy product* of $f(x)$ and $g(x)$ as

$$h(x) = \sum_{n=0}^{\infty} c_n x^n,$$

where

$$c_n = \sum_{k=0}^{n} a_k b_{n-k} = \sum_{k=0}^{n} a_{n-k} b_k.$$

(Thus, h is obtained by multiplying the series $\sum a_n x^n$ and $\sum b_n x^n$ as if they are ordinary polynomials.) Show that $\sum_{n=0}^{\infty} c_n x^n$ converges to $f(x)g(x)$ for $|x| < r$.

REVIEW PROBLEMS FOR CHAPTER 7

1. Let $f_n(x) = (2 + x^n)/(3 + x^n)$, $0 \le x < 1$. Show that the sequence $\{f_n\}$ converges pointwise, and determine whether the sequence converges uniformly on $[0, 1)$.
2. Give an example of a uniformly convergent series $\sum_{n=1}^{\infty} f_n$ of real-valued functions for which the Weierstrass M–test fails; in other words, if for each n we have $|f_n(x)| \le M_n$ for all x, then $\sum_{n=1}^{\infty} M_n = \infty$.
3. Let $f_n(x) = e^{-(x-n)}$ if $x \ge n$; $f_n(x) = 0$ if $x < n$. Show that the sequence converges pointwise on R, and determine whether the sequence converges uniformly.

4. From the power series expansion

$$\frac{1}{1+x} = 1 - x + x^2 - x^3 + \cdots \qquad \text{for} \qquad -1 < x < 1,$$

we obtain, by integrating term by term from 0 to x,

$$\ln(1+x) = x - \frac{x^2}{2} + \frac{x^3}{3} - \frac{x^4}{4} + \cdots \qquad \text{for} \qquad -1 < x < 1.$$

Indicate why the term-by-term integration is justified.

8

FURTHER TOPOLOGICAL RESULTS

8.1 THE EXTENSION PROBLEM

In this chapter we consider a variety of questions related to the topology of metric spaces. First, we examine the problem of extending a bounded continuous function from a closed subset to the entire space. The following result essentially solves the problem.

8.1.1 THEOREM. *Let A and B be disjoint closed subsets of the metric space Ω. There is a continuous $f:\Omega \to [0,1]$ such that $f = 0$ on A and $f = 1$ on B.*

This result holds for a wider class of topological spaces ("normal spaces"), and the general result is known as *Urysohn's Lemma*.

Proof. If $C \subseteq \Omega$, let $d(x,C)$ be the distance from x to C; that is, $d(x,C) = \inf_{y\in C} d(x,y)$. As shown in Section 4.4.2, the mapping $x \to d(x,C)$ is uniformly continuous on Ω. If $x \in C$, then $d(x,C) = 0$; if $x \notin C$ and C is closed, then $d(x,C) > 0$. Thus, if

$$f(x) = \frac{d(x,A)}{d(x,A) + d(x,B)},$$

f has the desired properties. ■

Note that if $g(x) = (b - a)f(x) + a$, then g is a continuous mapping of Ω into $[a, b]$, and $g = a$ on A, $g = b$ on B.

Before getting to the extension question, we must raise an issue that we have avoided so far, that of *relative topology*. If Ω is a metric space and $A \subseteq E \subseteq \Omega$, we know what it means to say that A is an open subset of Ω: if $x \in A$, there is an $r > 0$ such that $\{y \in \Omega; d(x,y) < r\} \subseteq A$. But A is also a subset of E, and if we regard E as the whole space, we then have a definition of A being open in E: if $x \in A$, there is an $r > 0$ such that $\{y \in E : d(x,y) < r\} \subseteq A$. Thus, if A is open in Ω, A must be open in E. The converse is not true, however. For example, $A = \{(x,y) : y = 0, a < x < b\}$ is open in $\{(x,y) : y = 0, -\infty < x < \infty\}$ (this just says, essentially, that an open interval is an open subset of R), but not open in R^2. For if $a < x < b$ and $y = 0$, no open ball centered at (x,y) is a subset of A.

Similarly, if $B \subseteq E \subseteq \Omega$, B is closed in Ω if $x_n \in B$, $x_n \to x \in \Omega$ implies $x \in B$; B is closed in E if $x_n \in B$, $x_n \to x \in E$ implies $x \in B$. Thus, if B is closed in Ω, then B is closed in E, but again not conversely. For example, $(0, 1]$ is closed in $(0, \infty)$ but not in R.

Under certain conditions, a set open (or closed) in the smaller space will be open (or closed) in the larger space.

8.1.2 THEOREM

(a) If A is open in E and E is open in Ω, then A is open in Ω.
(b) If B is closed in E and E is closed in Ω, then B is closed in Ω.

Proof

(a) If $x \in A$, there is an $r > 0$ such that $\{y \in E : d(x,y) < r\} \subseteq A$. Since E is open in Ω, there is an $s > 0$ such that $\{y \in \Omega : d(x,y) < s\} \subseteq E$. Thus, $\{y \in \Omega : d(x,y) < \min(r,s)\} \subseteq A \cap E = A$.
(b) Let $x_n \in B$, $x_n \to x \in \Omega$. Since $B \subseteq E$ and E is closed, $x \in E$. Since B is closed in E, $x \in B$. ■

We may now attack the extension problem.

8.1.3 Tietze Extension Theorem

Let $f: E \to R$, where E is a closed subset of the metric space Ω, and f is bounded and continuous on E. There is a continuous $g: \Omega \to R$ such that $g = f$ on E and $\sup\{|g(x)| : x \in \Omega\} = \sup\{|f(x)| : x \in E\}$. *Thus, f can be extended to the entire space without increasing the bound.*

Proof. Let $M = \sup_{x \in E} |f(x)| < \infty$, and define

$$A_1 = \{x \in E : f(x) \le -M/3\}, \qquad B_1 = \{x \in E : f(x) \ge M/3\}.$$

By continuity of f, A_1 and B_1 are closed in E, and hence in Ω by Theorem 8.1.2(b). By Theorem 8.1.1 there is a continuous function $h_1 : \Omega \to [-\frac{1}{3}M, \frac{1}{3}M]$ such that $h_1 = -\frac{1}{3}M$ on A_1, $h_1 = \frac{1}{3}M$ on B_1.

Now $f_2 = f - h_1$ is continuous on E. On A_1, $f \le -\frac{1}{3}M$ and $h_1 = -\frac{1}{3}M$; on B_1, $f \ge \frac{1}{3}M$ and $h_1 = \frac{1}{3}M$; on $A_1^c \cap B_1^c$, $-\frac{1}{3}M < f < \frac{1}{3}M$ and $-\frac{1}{3}M \le h_1 \le \frac{1}{3}M$. Thus, $|f_2| \le \frac{2}{3}M$ on E.

Continuing, we define

$$A_2 = \left\{x \in E : f_2(x) \le -\frac{1}{3}\left(\frac{2}{3}M\right)\right\},$$

$$B_2 = \left\{x \in E : f_2(x) \ge \frac{1}{3}\left(\frac{2}{3}M\right)\right\}.$$

Just as above, we find a continuous $h_2 : \Omega \to [-\frac{1}{3}(\frac{2}{3}M), \frac{1}{3}(\frac{2}{3}M)]$ such that $h_2 = -\frac{1}{3}(\frac{2}{3}M)$ on A_2, $h_2 = \frac{1}{3}(\frac{2}{3}M)$ on B_2. If $f_3 = f_2 - h_2 = f - h_1 - h_2$, then f_3 is continuous on E and $|f_3| \le \frac{2}{3}(\frac{2}{3}M) = (\frac{2}{3})^2 M$.

Inductively, we obtain continuous real-valued functions h_n on Ω such that $|h_n| \le \frac{1}{3}(\frac{2}{3})^{n-1}M$ on Ω and

$$\left| f - \sum_{j=1}^{n} h_j \right| \le \left(\frac{2}{3}\right)^n M \qquad \text{on} \quad E.$$

By the Weierstrass M–test, $\sum_{n=1}^{\infty} h_n$ converges uniformly to a continuous function g on Ω, where $|g| \le \frac{1}{3}M \sum_{n=1}^{\infty}(\frac{2}{3})^{n-1} = M$ and $f = g$ on E. ■

Problems for Section 8.1

1. If A and F are subsets of E, show that if A is open in E then $A \cap F$ is open in F.
2. If A and F are subsets of E and A is closed in E, show that $A \cap F$ is closed in F.
3. Let $f: E \to R$, where E is a closed subset of the metric space Ω. (Here we do *not* assume f bounded, as in Theorem 8.1.3.) Show that there is a continuous $g : \Omega \to \bar{R}$ such that $g = f$ on E.

(One way to proceed is to find a *homeomorphism*, i.e., a one–to–one, onto, continuous map h with a continuous inverse, mapping $\bar{R}$ to a closed, bounded interval of R, and apply Theorem 8.1.3. Another application of this idea is to make $\bar{R}$ into a metric space via $\bar{d}(x,y) = |h(x) - h(y)|$.)

8.2 BAIRE CATEGORY THEOREM

We now turn to a different question. Suppose we wish to construct a function $f: R \to R$ that is continuous at each irrational point and discontinuous at each rational point. This can be done explicitly; if $r_1, r_2, \ldots$ is an enumeration of the rationals, let $f(r_n) = a_n$, where $a_n > 0$ and $a_n \to 0$ as $n \to \infty$. If x is irrational, define $f(x) = 0$. Now if x is an irrational number and $\epsilon > 0$, we have $0 < a_n < \epsilon$ for all sufficiently large n, say for $n \geq N$. Choose $\delta > 0$ so small that none of the rational numbers r_n, $n < N$, belong to $(x - \delta, x + \delta)$. If $y \in (x - \delta, x + \delta)$, then either y is irrational or $y = r_n$ for some $n \geq N$. In any event, $0 \leq f(y) < \epsilon$. Thus, f is continuous at x. If x is a rational r_k, let $x_1, x_2, \ldots$ be a sequence of irrationals converging to x. Then $f(x_n) = 0$ for all n, but $f(x) = a_k > 0$. Therefore, f is discontinuous at x, as desired.

Perhaps surprisingly, it is not possible to find a function $f: R \to R$ that is continuous at each rational point and discontinuous on the irrationals. This is one of the consequences of the Baire Category Theorem, which we are about to consider. The theorem is concerned with sets that might be called "thin," namely sets that are nowhere dense.

8.2.1 Definitions and Comments

If A is a subset of the metric space Ω, A is said to be *nowhere dense* if its closure $\bar{A}$ has empty interior. (In general, the interior of a set B, denoted by B^0, is defined as the union of all open subsets of B. Thus, for A to be nowhere dense, there can be no open subsets of $\bar{A}$ except $\emptyset$. We should point out that for our present purposes, "open" always means "open in Ω.")

For example, any finite set is nowhere dense. Also, we saw in Theorem 4.4.1(e) that the Cantor set is nowhere dense. Note that the rationals Q do not form a nowhere dense subset of R, since $\bar{Q} = R$.

It follows from the definition that A is nowhere dense if and only if $(\bar{A})^c$ is dense. For if A is not nowhere dense, let V be a nonempty open subset of $\bar{A}$. If $x \in V$, no sequence in $(\bar{A})^c$ can converge to x, for if so the sequence would eventually be in V and hence in $\bar{A}$. Thus, $(\bar{A})^c$ is not dense. On the other hand, if A is nowhere dense, $x \in \Omega$, and $r > 0$, the nonempty open set $B_r(x)$ cannot be contained in $\bar{A}$. Take $r = 1/n$ to obtain points x_n with $d(x, x_n) < 1/n$ and $x_n \notin \bar{A}$. Then $x_n \to x$, proving $(\bar{A})^c$ dense.

To summarize: A is not nowhere dense iff there is an open ball $B_r(x) \subseteq \bar{A}$ iff there is an x that cannot be approximated arbitrarily closely by points in $(\bar{A})^c$ iff $(\bar{A})^c$ is not dense.

The set $B \subseteq \Omega$ is said to be of *category* 1 in Ω if B can be expressed as a countable union of nowhere dense subsets of Ω. Otherwise B is said to be of *category* 2 in Ω.

Now, the main result.

8.2.2 Baire Category Theorem

Let Ω be a nonempty, complete metric space. If $\Omega = \bigcup_{n=1}^{\infty} A_n$, where the A_n are closed subsets of Ω, then the interior of A_n is nonempty for some n. Therefore, Ω is of category 2 in itself.

Proof. Assume that the interior of A_n is empty for every n. The key idea is to find a sequence of open balls $B_{\delta_n}(x_n)$ satisfying $0 <$

$\delta_n < (\frac{1}{2})^n$ and $B_{\delta_n}(x_n) \subseteq B_{\delta_{n-1}/2}(x_{n-1})\backslash A_n$ (hence $B_{\delta_n}(x_n) \cap A_n = \emptyset$). Suppose we can do this. Then $x_n \in B_{\delta_{n-1}/2}(x_{n-1})$, so $d(x_n, x_{n-1}) < \frac{1}{2}(\frac{1}{2})^{n-1} = (\frac{1}{2})^n$; consequently (if $n < m$),

$$\begin{aligned} d(x_n, x_m) &\le d(x_n, x_{n+1}) + d(x_{n+1}, x_{n+2}) + \cdots + d(x_{m-1}, x_m) \\ &< (\tfrac{1}{2})^{n+1} + \cdots + (\tfrac{1}{2})^m \\ &< (\tfrac{1}{2})^n \to 0 \quad \text{as} \quad n, m \to \infty. \end{aligned}$$

Thus, $\{x_n\}$ is a Cauchy sequence. By the completeness hypothesis, x_n converges to a limit $x \in \Omega$. Now if $k > n$,

$$\begin{aligned} B_{\delta_k}(x_k) &\subseteq B_{\delta_{k-1}/2}(x_{k-1}) \subseteq B_{\delta_{k-1}}(x_{k-1}) \subseteq B_{\delta_{k-2}/2}(x_{k-2}) \subseteq \cdots \\ &\subseteq B_{\delta_n/2}(x_n). \end{aligned}$$

Thus, $x_k \in B_{\delta_n/2}(x_n)$ for all $k > n$; hence, $d(x, x_n) \le \frac{1}{2}\delta_n < \delta_n$. We conclude that $x \in B_{\delta_n}(x_n)$ for all n, so for all n, $x \notin A_n$. This contradicts the assumption that $\bigcup_{n=1}^{\infty} A_n = \Omega$.

To obtain the required sequence of open balls, note that since $A_1^0 = \emptyset$, we must have $A_1 \ne \Omega$ (because Ω is a nonempty open subset of itself). Therefore, A_1^c is a nonempty open set, so there is an open ball $B_{\delta_1}(x_1) \subseteq A_1^c$ with $0 < \delta_1 < \frac{1}{2}$. Since $A_2^0 = \emptyset$, $B_{\delta_1/2}(x_1)$ is not a subset of A_2; hence, $B_{\delta_1/2}(x_1)\backslash A_2$ is a nonempty open set and, therefore, contains an open ball $B_{\delta_2}(x_2)$ with $0 < \delta_2 < \frac{1}{4}$ (see Fig. 8.2.1). Similarly, $B_{\delta_2/2}(x_2)$ is not a subset of A_3, so we obtain $B_{\delta_3}(x_3) \subseteq B_{\delta_2/2}(x_2)\backslash A_3$, where $0 < \delta_3 < \frac{1}{8}$. Continuing inductively, we obtain the desired sequence. ■

If U_n is an open dense subset of the metric space Ω, then U_n^c is closed and $(U_n^c)^c = U_n$ is dense; hence, U_n^c is nowhere dense. If Ω is nonempty and complete, then by Theorem 8.2.2, $\bigcup_{n=1}^{\infty} U_n^c \ne \Omega$; that is, $\bigcap_{n=1}^{\infty} U_n \ne \emptyset$. In fact, a stronger result holds.

8.2.3 THEOREM. *If U_n is an open dense subset of the complete metric space Ω ($n = 1, 2, \ldots$), then $\bigcap_{n=1}^{\infty} U_n$ is dense.*

Proof. There is nothing to prove if Ω is empty, so assume Ω is nonempty. Let B be any open ball; the closure $\bar{B}$ is a closed

Figure 8.2.1 Proof of the Baire Category Theorem

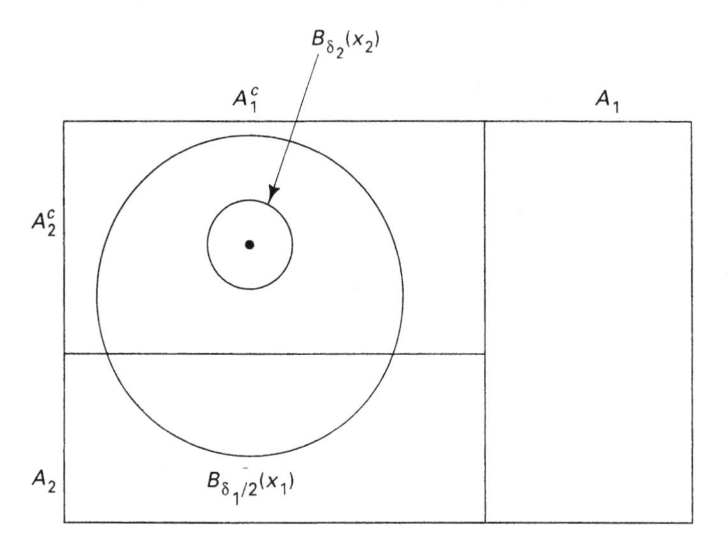

set and therefore is a complete metric space. (If $\{x_n\}$ is a Cauchy sequence in $\bar{B}$, then x_n converges to a limit $x \in \Omega$ and, necessarily, $x \in \bar{B}$ because $\bar{B}$ is closed.)

We claim that $U_n \cap B$ is dense in $\bar{B}$. For if $x \in \bar{B}$, we can find $y_n \in B$ with $d(x, y_n) < 1/2n$, and then find $z_n \in U_n \cap B$ with $d(y_n, z_n) < 1/2n$ (since U_n is dense in Ω). Thus, $d(x, z_n) < 1/n \to 0$, proving $U_n \cap B$ dense in $\bar{B}$. Also, $U_n \cap B$ is open in Ω and therefore open in $\bar{B}$. By the remarks before the statement of the theorem, $(\bigcap_{n=1}^{\infty} U_n) \cap B \neq \emptyset$. But then $\bigcap_{n=1}^{\infty} U_n$ intersects every open ball, so if $x \in \Omega$ we can find $x_j \in B_{1/j}(x) \cap \bigcap_{n=1}^{\infty} U_n$. Thus, $x_j \in \bigcap_{n=1}^{\infty} U_n$, $x_j \to x$. We conclude that $\bigcap_{n=1}^{\infty} U_n$ is dense. ■

The following result is another direct consequence of the Baire category theorem.

8.2.4 THEOREM. *The set of irrational numbers cannot be expressed as a countable union of closed sets; hence, the rationals cannot be expressed as a countable intersection of open sets.*

Proof. Let I be the set of irrationals and Q the set of rationals. If $I = \bigcup_{n=1}^{\infty} C_n$, with all C_n closed (in R), then each C_n is nowhere dense. To see this, note that any nonempty open subset of C_n must contain an open interval and therefore must contain rational numbers, contradicting $C_n \subseteq I$. Thus, C_n has empty interior. Hence, I is a countable union of nowhere dense sets, as is Q (if $\{r_n\}$ is an enumeration of the rationals, $Q = \bigcup_{n=1}^{\infty}\{r_n\}$). Therefore, R is a countable union of nowhere dense sets, contradicting Theorem 8.2.2. ■

A countable union of closed sets is often referred to as an F_σ *set*; a countable intersection of open sets is called a G_δ *set*.

The problem of finding a function continuous on the rationals and discontinuous on the irrationals may now be attacked. The following way of expressing the idea of continuity at a point will be helpful.

8.2.5 LEMMA. *Let $f:\Omega \to \Omega'$, where Ω and Ω' are metric spaces. If x is a point of Ω, then f is continuous at x if and only if for every positive integer n there is a $\delta > 0$ such that if $x_1, x_2 \in \Omega$ and $d(x_1,x) < \delta$, $d(x_2,x) < \delta$, we have $d(f(x_1), f(x_2)) < 1/n$.*

Proof. If f is continuous at x, let us choose $\delta > 0$ so that $d(y,x) < \delta$ implies $d(f(y), f(x)) < 1/2n$. If $d(x_1,x) < \delta$ and $d(x_2,x) < \delta$, then $d(f(x_1), f(x_2)) \le d(f(x_1), f(x)) + d(f(x), f(x_2)) < 1/n$. Conversely, if the given condition holds, take $x_2 = x$ to show continuity at x. ■

The key result is the following.

8.2.6 THEOREM. *Let f be an arbitrary mapping from the metric space Ω to the metric space Ω'. Then $\{x \in \Omega : f$ is discontinuous at $x\}$ is an F_σ, that is, a countable union of closed sets. Thus, by Theorem 8.2.4, there is no $f: R \to R$ that is continuous at each rational point and discontinuous at each irrational point.*

Proof. By Lemma 8.2.5, f is continuous at x if and only if

$$(\forall n)(\exists \delta > 0)(\forall x_1, x_2 \in \Omega)(d(x_1,x) < \delta,$$
$$d(x_2,x) < \delta \Rightarrow d(f(x_1), f(x_2)) < \frac{1}{n}).$$

Thus, f is discontinuous at x if and only if $(\exists n)(\forall \delta > 0)(\exists x_1, x_2 \in \Omega)$ such that

$$d(x_1,x) < \delta, \quad d(x_2,x) < \delta, \quad \text{and} \quad d(f(x_1), f(x_2) \ge 1/n.$$

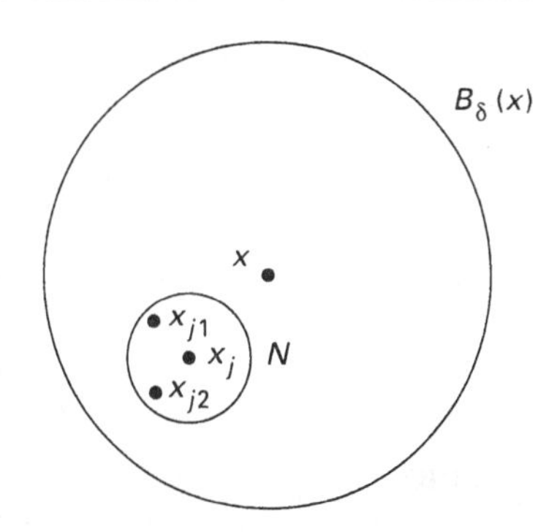

Figure 8.2.2 Proof of Theorem 8.2.6

Therefore, $\{x : f \text{ is discontinuous at } x\} = \bigcup_{n=1}^{\infty} D_n$, where

$$D_n = \{x : \forall\delta > 0\ \exists x_1, x_2 \in \Omega \quad \text{such that} \\ d(x_1, x) < \delta, d(x_2, x) < \delta, \quad \text{and} \quad d(f(x_1), f(x_2)) \geq 1/n\}.$$

We are finished if we can show that the D_n are closed. Let $x_j \in D_n$, $x_j \to x$. Given $\delta > 0$, we have $x_j \in B_\delta(x)$ for all sufficiently large j. Pick any such j, and let N be an open ball centered at x_j and entirely contained in $B_\delta(x)$, say $N = B_{\delta'}(x_j)$ (see Fig. 8.2.2). Since $x_j \in D_n$, there are points $x_{j1}, x_{j2} \in N$ with $d(f(x_{j1}), f(x_{j2})) \geq 1/n$. But then (since $N \subseteq B_\delta(x)$, so that $d(x_{j1}, x)$ and $d(x_{j2}, x)$ are less than δ) $x \in D_n$, proving D_n closed. ■

Problems for Section 8.2

1. If $B_r(x)$ is an open ball in a metric space, show that

$$\overline{B_r(x)} \subseteq \{y : d(x,y) \leq r\}.$$

2. Give an example of a metric space for which the inclusion in Problem 1 is proper; i.e., verify that $\{y : d(x,y) \leq r\}$ can be strictly larger than $\overline{B_r(x)}$.

3. Let $f_1, f_2, \ldots$ be real-valued functions on the metric space Ω. Show that

$$\{x \in \Omega : f_n(x) \to f(x) \quad \text{as} \quad n \to \infty\} = \\ \bigcap_{m=1}^{\infty} \bigcup_{n=1}^{\infty} \bigcap_{k=n}^{\infty} \left\{x \in \Omega : |f_k(x) - f(x)| < \frac{1}{m}\right\}.$$

8.3 CONNECTEDNESS

Often in analysis, sets occur that consist of several "parts" or "pieces." One solves a problem on each piece separately and then puts the results together. A topological question that is of interest is "When is a set in one piece?" For example, the set $E = (0,1) \cup (1,2)$ seems to have two parts, and one observation that can be made is that $(0,1)$ and $(1,2)$ are each open and closed in E. (Note that $(0,1)$ is not closed in R since, for example, $1 - 1/n \to 1 \notin (0,1)$. But $1 \notin E$, and in fact there is no sequence in $(0,1)$ converging to a point of E outside $(0,1)$. Thus $(0,1)$ is closed in E.) This idea is used to define a general notion of connectedness.

8.3.1 Definitions

Let E be a subset of the metric space Ω. We say that E is *disconnected* if E can be expressed as $A \cup B$, where A and B are disjoint, nonempty, and both open (equivalently, both closed) in E. A set is called *connected* if it is not disconnected.

A somewhat more natural notion of connectedness is the following.

The set E is *path–connected* if for all $x, y \in E$, there is a path in E joining x to y; that is, there is a continuous $f\colon [0,1] \to E$ such that $f(0) = x$, $f(1) = y$. If we think of t as a time variable, we are at the point x at time $t = 0$, at y at time $t = 1$, and we move smoothly in E from x to y as time goes by.

Path–connectedness is stronger than connectedness, as we now show.

8.3.2 THEOREM. *If E is path–connected, then E is connected. Thus, all intervals of R^p are connected, as is R^p itself.* (An interval of R^p is defined in the natural way, for example, if $a = (a_1, \ldots, a_p)$, $b = (b_1, \ldots, b_p)$, then

$$(a,b] = \{(x_1 \ldots, x_p) \in R^p : a_i < x_i \le b_i,\ 1 \le i \le p\}.)$$

Proof. Assume $E = A \cup B$, where A and B are disjoint, nonempty, and closed in E. Pick $x \in A$, $y \in B$, and let $f\colon [0,1] \to E$ be continuous, with $f(0) = x$, $f(1) = y$. Let $S = \{t \in [0,1] : f(t) \in A\}$, and let $t_0 = \sup S$; note $t_0 > 0$ because A is open in E. Since $t_0 - 1/n$

is not an upper bound of S, we can find $t_n \in S$ with $t_0 - 1/n < t_n \le t_0$. Thus, $t_n \to t_0$ and $f(t_n) \in A$; since f is continuous and A is closed in E, we have $f(t_n) \to f(t_0) \in A$. Now $t_0 < 1$ because $f(1) = y \in B$, and since t_0 is an upper bound of S, $f(t) \in B$ for all $t > t_0$ (with $t \in [0, 1]$). Consider any sequence of points t_n' approaching t_0 from above. Since f is continuous and B is closed in E, $f(t_n') \to f(t_0) \in B$. This contradicts the disjointness of A and B. ■

An example of a set that is connected but not path–connected is the "topologist's sine curve" $E = F \cup G$, where $F = \{(x,y) \in R : y = \sin 1/x, 0 < x \le 1, \}, G = \{(0,0)\}$.

The following result is often useful in proving connectedness.

8.3.3 THEOREM. *Let E be the union of the connected sets F and G, and assume that $\bar{F} \cap G \neq \emptyset$ or $F \cap \bar{G} \neq \emptyset$. Then E is connected.*

Proof. We may as well assume $\bar{F} \cap G \neq \emptyset$. Suppose A is open and closed in E. Then $A \cap F$ is open and closed in F (Section 8.1, Problem 1), so $A \cap F = \emptyset$ or F. Similarly, $A \cap G = \emptyset$ or G.

Case 1. $A \cap F = \emptyset, A \cap G = \emptyset$. Then $A = A \cap E = (A \cap F) \cup (A \cap G) = \emptyset$.

Case 2. $A \cap F = F, A \cap G = G$. Then $F \subseteq A, G \subseteq A$; hence $E = F \cup G \subseteq A$, so $A = E$.

Case 3. $A \cap F = F, A \cap G = \emptyset$. Let $z \in \bar{F} \cap G$, and let $z_n \in F$, $z_n \to z$. Then $z_n \in A$, $z_n \to z \in G$ (hence $z \notin A$), contradicting the fact that A is closed in E.

Case 4. $A \cap F = \emptyset, A \cap G = G$. If $A_0 = E \backslash A$, then (since $G \subseteq A$) $A_0 \cap G = \emptyset$, and $A_0 \cap F = F$. (Note that $A \cap F = \emptyset$ implies $F \subseteq A^c$, and, by hypothesis, $F \subseteq E$. Thus $F \subseteq A_0$.) The situation is now the same as in Case 3, and a contradiction results.

We conclude that $A = \emptyset$ or $A = E$, proving E connected. ■

Problems for Section 8.3

1. The subsets A and B of the metric space Ω are said to be *separated* iff A and B are disjoint and both closed (hence open) in $A \cup B$.

Thus, E is disconnected iff E can be expressed as the union of two nonempty separated sets. Show that A and B are separated if and only if $\bar{A} \cap B = A \cap \bar{B} = \emptyset$.

2. Continuing Problem 1, show that A and B are separated if and only if there exist open sets G_1, G_2, with $A \subseteq G_1, B \subseteq G_2$, and $G_1 \cap B = A \cap G_2 = \emptyset$.

(The conditions given in Problems 1 and 2 are useful because they express the idea of connectedness without using the relative topology.)

3. Let E be a connected subset of R, and let $a = \inf E, b = \sup E$. If $a < x < b$, show that $x \in E$. Thus, the only connected subsets of R are the intervals.
4. If f is a continuous function on the connected set E, show that the image $f(E)$ is connected.
5. Use Problems 3 and 4 to give an alternative proof of the Intermediate Value Theorem 4.3.2.

8.4 SEMICONTINUOUS FUNCTIONS

A standard example of a sequence of continuous functions f_n converging pointwise to a discontinuous limit f is

$$f_n(x) = \begin{cases} 0, & x \leq 0, \\ nx, & 0 < x \leq 1/n, \\ 1, & x > 1/n, \end{cases}$$

$$f(x) = \begin{cases} 0, & x \leq 0, \\ 1, & x > 0. \end{cases}$$

Although the limit function is discontinuous, it does have a "semicontinuity" property, as follows.

8.4.1 Definitions and Comments

Let $f \colon \Omega \to \bar{R}$, where Ω is a metric space. We say that f is *lower semicontinuous* (LSC) on Ω if for every $a \in \bar{R}$, $\{x \in \Omega : f(x) > a\}$ is open; f is *upper semicontinuous* (USC) on Ω if for every $a \in \bar{R}$, $\{x \in \Omega : f(x) < a\}$ is open.

Possibly a useful way of remembering the definition is to note that the function f given above is LSC on R; it takes the lower value at the discontinuity. Also, in the set $\{x : f(x) > a\}$ involved in the definition of lower semicontinuity, a is the lower number in the inequality. Similarly, if $g(x) = 0, x < 0; g(x) = 1, x \geq 0$, then g is USC on R. It takes the upper value at the discontinuity, and in the set $\{x : g(x) < a\}$ involved in the definition of upper semicontinuity, a is the upper number in the inequality.

A function $f: \Omega \to \bar{R}$ is continuous on Ω if and only if it is both USC and LSC. To see this, note that if f is USC and LSC, then for all $a\ b \in \bar{R}$, $a < b$, the set $\{x : a < f(x) < b\}$ is open. If V is an arbitrary open subset of $\bar{R}$, then V is a union of open intervals I_n; hence, $f^{-1}(V) = \bigcup_n f^{-1}(I_n)$ is open.

In this argument we have glossed over (except in Section 8.1, Problem 3) the fact that we do not yet have a metric on $\bar{R}$. (If we use ordinary Euclidean distance, we get $d(x, \infty) = \infty$ for every $x \in R$, which is awkward.) The easiest way out is to identify $\bar{R}$ with a closed bounded interval of reals, for example, with the interval $[0, \pi]$ by means of the one–to–one onto mapping $h(x) = \pi/2 + \arctan x, x \in \bar{R}$. If $x, y \in \bar{R}$, define $d(x, y)$ to be $|h(x) - h(y)|$. Then V will be open in $\bar{R}$ if and only if $h(V)$ is open in $[0, \pi]$. In particular, $(a, \infty]$ and $[-\infty, a)$ are open in $\bar{R}$.

Note that f is LSC if and only if $-f$ is USC; this often allows proofs about LSC functions to apply to USC functions as well.

The following criterion for semicontinuity is useful.

8.4.2 THEOREM. *Let $f: \Omega \to \bar{R}$. Then f is LSC on Ω if and only if for each $x \in \Omega$ and every sequence $\{x_n\}$ in Ω with $x_n \to x$, we have* $\lim_n \inf f(x_n) \geq f(x)$. *Similarly, f is USC on Ω if and only if $x_n \to x$ implies* $\lim_n \sup f(x_n) \leq f(x)$.

Proof. Let f be LSC. If $x_n \to x$ and b is any number less than $f(x)$, then $V = \{y \in \Omega : f(y) > b\}$ is open, and $x \in V$. For large enough n, $x_n \in V$; hence $f(x_n) > b$. But this implies that any convergent subsequence of $\{f(x_n)\}$ has a limit that is at least b. We conclude that $\lim_n \inf f(x_n) \geq b$. But b may be chosen arbitrarily close to $f(x)$, and therefore $\lim_n \inf f(x_n) \geq f(x)$.

Conversely, assume that $x_n \to x$ implies $\lim_n \inf f(x_n) \geq f(x)$, and let $W = \{x : f(x) > a\}$. We show that W^c is closed (hence W is open). Let $x_n \in W^c$, $x_n \to x$. Then $f(x_n) \leq a$ for all n, and $\lim_n \inf f(x_n) \geq f(x)$ by assumption. Therefore $f(x) \leq a$; that is, $x \in W^c$. Thus W^c is closed.

The USC result may be proved in a similar fashion or by applying the LSC result to $-f$. ■

The condition of Theorem 8.4.2 may be used, if desired, to define upper and lower semicontinuity at a point $x \in \Omega$.

The next result is a direct consequence of the definition of semicontinuity.

8.4.3 THEOREM. *If $\{f_i, i \in I\}$ is an arbitrary family of LSC functions, then* $\sup_{i\in I} f_i$ *is LSC. If $f_1, \ldots, f_n$ are LSC, then* $\min_{1\leq i\leq n} f_i$ *is LSC.*

Similarly, the inf *of an arbitrary family of USC functions is USC, and the maximum of a finite number of USC functions is USC.*

(All of the above functions are defined pointwise; for example, if $h = \sup_{i\in I} f_i$, then for each x we have $h(x) = \sup\{f_i(x) : i \in I\}$.)

Proof. Assume f_i LSC for each i, and let $f = \sup_i f_i$. If $a \in \bar{R}$, then $\{x: f(x) > a\} = \bigcup_i \{x: f_i(x) > a\}$, which is open. If $g = \min_{1\leq i\leq n} f_i$, then $\{x: g(x) > a\} = \bigcap_{i=1}^n \{x: f_i(x) > a\}$, again an open set. The proof for USC functions is analogous. ■

We have seen that a continuous real-valued function on a compact set has a maximum and a minimum. Part of this result is retained in the semicontinuous case.

8.4.4 THEOREM. *If f is LSC on the compact metric space Ω, then f attains a minimum; that is, for some $x \in \Omega$ we have $f(x) = \inf_{y\in\Omega} f(y)$.*

Similarly, an USC function on a compact metric space attains a maximum.

Proof. Assume f LSC. If $b = \inf_{y\in\Omega} f(y)$, there is a sequence of points $x_n \in \Omega$ such that $f(x_n) \to b$. Since Ω is compact, there is by Theorem 2.3.1 a subsequence x_{n_k} converging to a limit $x \in \Omega$. By Theorem 8.4.2, $\lim_k \inf f(x_{n_k}) \geq f(x)$, and therefore $b \geq f(x)$. But $f(x) \geq b$ by definition of b, so $f(x) = b$. The USC case is handled similarly. ■

If f has a jump at x and, say, $f(x^-) < f(x) < f(x^+)$, then, intuitively, $f(x^-)$ looks like the lim inf of $f(y)$ as y approaches x. In other words, among all possible limits of sequences $f(x_n)$ as $x_n \to x$, $f(x^-)$ is the smallest. Similarly, $f(x^+)$ looks like the lim sup of $f(y)$ as $y \to x$. The formalization of these ideas leads to the discovery of the largest LSC function that is less than or equal to a given function f, and the smallest USC function greater than or equal to f.

8.4.5 THEOREM. *Let $f\colon \Omega \to \bar{R}$ be arbitrary. If $x \in \Omega$, let $N(x)$ denote the class of open sets containing x (also called the class of neighborhoods of x). Define*

$$\liminf_{y \to x} f(y) = \sup_{V \in N(x)} \inf_{y \in V} f(y),$$

$$\limsup_{y \to x} f(y) = \inf_{V \in N(x)} \sup_{y \in V} f(y).$$

(Note the analogy with the formulas $\liminf_{n\to\infty} x_n = \sup_n \inf_{k \geq n} x_k$, $\limsup_{n\to\infty} x_n = \inf_n \sup_{k \geq n} x_k$. Intuitively, $\inf_{y \in V_1} f(y) \leq \inf_{y \in V_2} f(y)$ if $V_2 \subseteq V_1$, so in computing $\liminf_{y \to x} f(y)$ we can ignore V_1. So in a sense, the $V \in N(x)$ are approaching x.)

Define the lower and upper envelopes of f by

$$\underline{f}(x) = \liminf_{y \to x} f(y), \quad \bar{f}(x) = \limsup_{y \to x} f(y), \qquad x \in \Omega.$$

Then $\underline{f}$ is LSC on Ω and if g is any LSC function on Ω such that $g \leq f$, we have $g \leq \underline{f}$. Similarly, $\bar{f}$ is USC on Ω, and if g is any USC function on Ω with $g \geq f$, then $g \geq \bar{f}$. Thus, $\underline{f}$ is the largest LSC function $\leq f$, and $\bar{f}$ is the smallest USC function $\geq f$.

Proof. As usual we consider only the LSC case; to prove the USC result, use a similar argument or apply the LSC result to $-f$.

Let $x_n \to x$. We wish to show that $\lim_n \inf \underline{f}(x_n) \geq \underline{f}(x)$; by Theorem 8.4.2, $\underline{f}$ will be LSC on Ω.

Suppose $\lim_n \inf \underline{f}(x_n) < b < \underline{f}(x)$. If V is any neighborhood of x, then there is an $x_n \in V$ with $\underline{f}(x_n) < b$. (There is a subsequence with $\underline{f}(x_{n_k}) \to \lim_n \inf \underline{f}(x_n) < b$; since $x_{n_k} \to x$ we have $x_{n_k} \in V$ for large k. Note that n depends on V.) But V is an open set containing x_n, so

$$\underline{f}(x_n) = \sup_{W \in N(x_n)} \inf_{y \in W} f(y) \geq \inf_{y \in V} f(y),$$

so for every $V \in N(x)$, $\inf_{y \in V} f(y) < b$. Thus,

$$\underline{f}(x) = \sup_{V \in N(x)} \inf_{y \in V} f(y) \leq b < \underline{f}(x),$$

a contradiction.

Now let g LSC, $g \leq f$. We claim that $\liminf_{y \to x} g(y) \geq g(x)$. For if $\sup_{V \in N(x)} \inf_{y \in V} g(y) < g(x)$, then for some $b < g(x)$ we have $\inf_{y \in V} g(y) < b$ for all $V \in N(x)$. Take $V = B_{1/n}(x)$, and select $x_n \in B_{1/n}(x)$ with $g(x_n) < b$. Thus, $x_n \to x$, but $\lim_n \inf g(x_n) \leq b < g(x)$, contradicting Theorem 8.4.2. The proof is completed by the observation that

$$\underline{f}(x) = \liminf_{y \to x} f(y) \geq \liminf_{y \to x} g(y) \geq g(x). \quad \blacksquare$$

If $f_1, f_2, \ldots$ are continuous, extended real–valued functions on Ω and $f_1 \leq f_2 \leq \ldots$, then $f = \lim_{n \to \infty} f_n$ is LSC by Theorem 8.4.3 (note $f = \sup_n f_n$). Similarly, if the f_n decrease to f, then f is USC, since in this case $f = \inf_n f_n$. We now prove that any semicontinuous function can be expressed as a monotone limit of continuous functions. (We adopt the following convenient abbreviations. If $f_1 \leq f_2 \leq \ldots$ and

$f_n \to f$ pointwise on Ω, we write $f_n \uparrow f$. If $f_1 \geq f_2 \geq \ldots$ and $f_n \to f$, we write $f_n \downarrow f$.)

8.4.6 THEOREM. *Let f be LSC on the metric space Ω. Then there is a sequence of continuous functions $f_n: \Omega \to \bar{R}$ with $f_n \uparrow f$. Similarly, if f is USC, there is a sequence of continuous f_n with $f_n \downarrow f$. If $|f| \leq M < \infty$ on Ω, the f_n can be chosen so that $|f_n| \leq M$ for all n.*

Proof. Again, we consider only the LSC case. First assume f nonnegative and finite valued. Define

$$f_n(x) = \inf_{z \in \Omega} [f(z) + n\, d(x,z)].$$

If $x, y \in \Omega$, then $f(z) + n\, d(x,z) \leq f(z) + n\, d(y,z) + n\, d(x,y)$. Take the inf over z (first on the left side, then on the right side) to obtain $f_n(x) \leq f_n(y) + n\, d(x,y)$. By symmetry,

$$|f_n(x) - f_n(y)| \leq n\, d(x,y);$$

hence, f_n is uniformly continuous on Ω. Furthermore, since $f \geq 0$, we have $0 \leq f_n(x) \leq f(x) + n\ d(x,x) = f(x)$. By definition, f_n increases with n; we must show that $\lim_n f_n$ is actually f.

Given $\epsilon > 0$, by definition of $f_n(x)$ there is a point $z_n \in \Omega$ such that

$$\begin{aligned} (1) \qquad f_n(x) + \epsilon &> f(z_n) + n\ d(x, z_n) \\ &\geq n\ d(x, z_n) \qquad \text{since} \quad f \geq 0. \end{aligned}$$

But $f_n(x) + \epsilon \leq f(x) + \epsilon$; hence $d(x, z_n) \to 0$. Since f is LSC, Theorem 8.4.2 yields $\lim_n \inf f(z_n) \geq f(x)$; hence

$$(2) \qquad f(z_n) > f(x) - \epsilon \qquad \text{eventually.}$$

By (1) and (2),

$$f_n(x) > f(z_n) - \epsilon + n\ d(x, z_n) \geq f(z_n) - \epsilon > f(x) - 2\epsilon$$

for all sufficiently large n. Thus, $f_n(x) \to f(x)$.

If $|f| \leq M < \infty$, then $f + M$ is LSC, finite–valued, and nonnegative. If $0 \leq g_n \uparrow f + M$, then $f_n = g_n - M \uparrow f$ and $|f_n| \leq M$.

In general, observe that $h(x) = \pi/2 + \arctan x, x \in \bar{R}$, is a one–to–one onto mapping of $\bar{R}$ onto $[0, \pi]$, continuous, and having a continuous inverse (such a mapping is called a homeomorphism). Also, $x < y$ if and only if $h(x) < h(y)$, so that h is order–preserving. Let $f_0(x) = h(f(x))$; then $\{x : f_0(x) > a\} = \{x : f(x) > h^{-1}(a)\}$, so f_0 is LSC. Furthermore, f_0 is finite–valued and nonnegative. By what we have proved above, there is a sequence of continuous functions g_n such that $g_n(x) \uparrow f_0(x) = h(f(x)), x \in \Omega$. If $f_n(x) = h^{-1}(g_n(x))$, then $f_n(x) \uparrow h^{-1}(h(f(x))) = f(x)$, as desired. ∎

Problems for Section 8.4

1. Give an example of an infinite sequence of LSC functions f_n such that $\inf_n f_n$ is *not* LSC.
2. Give an example of a (nonmonotone) limit of continuous functions that is neither USC nor LSC.
3. Give an example of a LSC function on a compact set such that f has no maximum.
4. Formulate an epsilon-delta statement that is equivalent to the condition of Theorem 8.4.2, and prove your result.
5. In Theorem 8.4.6, if f is LSC and nonnegative, show that the functions f_n can also be taken to be nonnegative.

REVIEW PROBLEMS FOR CHAPTER 8

1. Let Ω be a metric space and let $A \subseteq E \subseteq \Omega$. If A is open in Ω, then A is open in E
   (a) always
   (b) sometimes
   (c) never

2. In the previous problem, if A is open in E then A is open in Ω
   (a) always
   (b) sometimes
   (c) never
3. Give an example of
   (a) a countably infinite nowhere dense subset of R.
   (b) an uncountably infinite nowhere dense subset of R.

9

EPILOGUE

9.1 SOME COMPACTNESS RESULTS

There are two major loose ends in our development of real analysis. We have never formally defined the real numbers, and when we developed the basic topology of Euclidean n–space, we assumed the Cantor nested set property (Section 2.2.2) without apology. We now attempt to face these problems; a convenient way to do this is via some general compactness results in metric spaces.

9.1.1 Definitions and Comments

Let A be a subset of the metric space Ω. We say that A is *totally bounded* if for every $\epsilon > 0$, A can be covered by finitely many open balls of radius ϵ.

It follows from the definition that any totally bounded set is bounded, but the converse is not true. For example, let Ω be any set, and let $d(x,y) = 1, x \neq y$; $d(x,x) = 0, x, y \in \Omega$. Then all subsets of Ω are bounded, but if Ω has infinitely many points, then Ω is not totally bounded. (If $0 < r < 1$, then $B_r(x) = \{x\}$.) In R^p, however, a bounded set must be totally bounded. (Intuitively, a bounded set can be placed inside a large rectangular box, which can be broken up into a finite number of small boxes of maximum dimension ϵ. Each of the small boxes can be covered by a finite number of open balls of radius ϵ.)

If A is totally bounded, then A has a countable dense subset E. To see this, note that for any positive integer n, A can be covered by finitely many balls $B_{1/n}(x_{nj})$, $j = 1, 2, \ldots, m_n$, with centers $x_{nj} \in A$. The set E consisting of all x_{nj} is countable and dense in A (see the proof of Lemma 7.4.1).

If every sequence in A has a convergent subsequence (the limit need not be in A), then A is totally bounded. For if A cannot be covered by finitely many open balls of radius ϵ, select, inductively, a sequence of points $x_n \in A$ with $x_2 \notin B_\epsilon(x_1)$, $x_3 \notin B_\epsilon(x_1) \cup B_\epsilon(x_2), \ldots, x_n \notin \bigcup_{i=1}^{n-1} B_\epsilon(x_i), \ldots$. If $m > n$, then $x_m \notin B_\epsilon(x_n)$; hence $d(x_m, x_n) \geq \epsilon$. Thus, $\{x_n\}$ has no convergent subsequence.

We may now give two general compactness criteria in metric spaces.

9.1.2 THEOREM. *Let K be a subset of the metric space Ω. Then K is compact if and only if every sequence in K has a subsequence converging to a point of K.*

Proof. The "only if" part is Theorem 2.3.1, so assume that every sequence in K has a subsequence converging to a point of K. As shown in Section 9.1.1, K is totally bounded and therefore has a countable dense subset. By Lemma 7.4.1, every open covering of K has a countable subcovering. Thus, we may start with the assumption that $K \subseteq \bigcup_{i=1}^{\infty} G_i$, where the G_i are open sets, and attempt to find a finite subcovering.

If for every n, K is not a subset of $\bigcup_{i=1}^{n} G_i$, select $x_n \in K$ with $x_n \notin \bigcup_{i=1}^{n} G_i$. By hypothesis there is a subsequence converging to a limit $x \in K$. Now x belongs to some G_j, and therefore the subsequence is in G_j eventually. Thus, $x_n \in G_j$ for some $n > j$, contradicting $x_n \notin \bigcup_{i=1}^{n} G_i$. ■

9.1.3 THEOREM. *Let K be a subset of the metric space Ω. Then K is compact if and only if K is totally bounded and complete. (Note that* *completeness of K means that every Cauchy sequence in K converges to a point in K.)*

Proof. If K is compact, then K is totally bounded by the definition of compactness. (Note that for each $\epsilon > 0$, $K \subseteq \bigcup_{x \in K} B_\epsilon(x)$, and by compactness there is a finite subcovering. Alternatively, Theorem 9.1.2 and Section 9.1.1 can be used, but this is a rather indirect approach.) Let $\{x_n\}$ be a Cauchy sequence in K; by Theorem 9.1.2 there is a subsequence $\{x_{n_k}\}$ converging to a point $x \in K$. We proceed as in the proof of Theorem 2.4.5. Given $\epsilon > 0$, choose N so that $d(x_n, x_m) < \epsilon/2$ for all $n, m \geq N$, and choose k_0 so that $n_k \geq N$ and $d(x_{n_k}, x) < \epsilon/2$ for all $k \geq k_0$. Then for any fixed $k \geq k_0$,

$$d(x_n, x) \leq d(x_n, x_{n_k}) + d(x_{n_k}, x) < \epsilon \qquad \text{for all} \quad n \geq N.$$

Thus, $x_n \to x \in K$.

Conversely, assume K totally bounded and complete, and let x_1, $x_2, \ldots$ be a sequence in K. By Theorem 9.1.2, it suffices to produce a subsequence converging to a point of K. Now for any r, K is covered by finitely many open balls of radius r; hence, for some ball $B = B_r(y)$ in the covering we must have $x_k \in B$ for infinitely many k. Using this idea successively for $r = 1/n$, $n = 1, 2, \ldots$, we obtain the following array, with row n a subsequence of row $n - 1$, $n = 2, 3, \ldots$, and row 1 a subsequence of $\{x_n\}$:

$$\begin{array}{cccccc}
x_{11} & x_{12} & x_{13} & \cdots & \in B_1(y_1) \\
x_{21} & x_{22} & x_{23} & \cdots & \in B_{1/2}(y_2) \\
 & & & \vdots & \\
x_{n1} & x_{n2} & x_{n3} & \cdots & \in B_{1/n}(y_n) \\
 & & & \vdots &
\end{array}$$

Let $z_n = x_{nn}$, $n = 1, 2, \ldots$. If $m \geq n$, then $z_m \in B_{1/n}(y_n)$; hence,

$$\begin{aligned}
d(z_m, z_n) &\leq d(z_m, y_n) + d(y_n, z_n) \\
&< \frac{1}{n} + d(y_n, x_{nn}) \\
&< \frac{1}{n} + \frac{1}{n} \to 0 \qquad \text{as} \quad n \to \infty.
\end{aligned}$$

Thus, $\{z_n\}$ is a Cauchy sequence, which by completeness converges to a point of K. ■

9.2 REPLACING CANTOR'S NESTED SET PROPERTY

We can now examine the problem of replacing our assumption 2.2.2 by something more natural. Suppose we assume instead that R^p is complete. The Heine–Borel Theorem can then be proved. For if K is a closed and bounded subset of R^p, then K is totally bounded by Section 9.1.1, and therefore compact by Theorem 9.1.3. (Note that K, a closed subset of the complete space R^p, must be complete.) From the Heine–Borel Theorem we can prove Cantor's nested set theorem, as follows. Let $B_1 \supseteq B_2 \supseteq \ldots$ be a nested sequence of closed, bounded, nonempty subsets of R^p, and pick $x_n \in B_n$, $n = 1, 2, \ldots$. Since B_1 is compact, there is by Theorem 9.1.2 a subsequence $\{x_{n_j}\}$ converging to a limit $x \in B_1$. For any positive integer k, we have $n_j \geq k$ for all sufficiently large j; hence $x_{n_j} \in B_{n_j} \subseteq B_k$. It follows that $x \in B_k$ for every k, as desired.

Now it follows directly from the Euclidean distance formula that a sequence of points $x_n = (x_{n1}, \ldots, x_{np}) \in R^p$ is Cauchy in R^p if and only if each of the component sequences $\{x_{nk}, n = 1, 2, \ldots\}$ is Cauchy in R. Similarly, convergence of $\{x_n\}$ is equivalent to convergence of each $\{x_{ni}\}$, $i = 1, \ldots, p$. Thus, it is sufficient to assume R complete. But in fact completeness of R follows from the assumption that every nonempty subset of R that has an upper bound has a supremum (equivalently, every nonempty subset of R that has a lower bound has an infimum). For under this assumption, let $\{x_n\}$ be a Cauchy sequence in R, and let $L = \inf_n \sup_{k \geq n} x_k$ (the inf and the sup must exist because $\{x_n\}$ is bounded; see Section 2.4.4).

We show that $x_n \to L$. If $\epsilon > 0$, then $\inf_n \sup_{k \geq n} x_k > L - \epsilon$; hence, for every n, $\sup_{k \geq n} x_k > L - \epsilon$. Thus,

$$(1) \qquad (\forall n)(\exists k \geq n)(x_k > L - \epsilon).$$

Also, $\inf_n \sup_{k \geq n} x_k < L + \epsilon$, so for some n, $\sup_{k \geq n} x_k < L + \epsilon$. Therefore,

$$(2) \qquad (\exists n)(\forall k \geq n)(x_k < L + \epsilon).$$

Now by the Cauchy property we have, for some N, $|x_m - x_n| < \epsilon$ for all $m, n \geq N$. By (2) there is an n_0 such that $x_k < L + \epsilon$ for all $k \geq n_0$, and by (1) we can find $k_0 \geq \max(n_0, N)$ such that $x_{k_0} > L - \epsilon$. Thus, $L - \epsilon < x_{k_0} < L + \epsilon$. Finally,

$$\begin{aligned} |x_n - L| &\leq |x_n - x_{k_0}| + |x_{k_0} - L| \\ &< 2\epsilon \qquad \text{for all} \quad n \geq N. \end{aligned}$$

Therefore $x_n \to L$.

9.3 THE REAL NUMBERS REVISITED

When the real numbers are introduced by means of a set of axioms, the standard assumptions are as follows.

1. R is a field.

Intuitively, this means that addition, subtraction, multiplication, and division can be carried out without leaving the set R.

2. R is an ordered field.

The idea is that there is an ordering on R, denoted by "$<$," such that if $y < z$, then $x + y < x + z$; and if $x > 0$ and $y > 0$, then $xy > 0$. Also for all x and y, exactly one of the three conditions $x < y, x = y$, $x > y$ holds.

3. R has the least upper bound property.

In other words, every nonempty subset of R that has an upper bound has a least upper bound.

Of course, we have glossed over a host of formal details, but this material belongs more to the areas of logic and algebra than to analysis. The student may consult a standard textbook to fill in the gaps.

SOLUTIONS TO PROBLEMS

SECTION 1.1

Note: "iff" stands for "if and only if"

1.
$$x \in \left(\bigcap_i A_i\right)^c \quad \text{iff } x \notin \left(\bigcap_i A_i\right) \text{ iff } x \notin (\text{all } A_i)$$
$$\text{iff for at least one } i, x \in A_i^c$$
$$\text{iff } x \in \bigcup_i A_i^c$$
$$x \in \left(\bigcup_i A_i\right)^c \quad \text{iff } x \notin \left(\bigcup_i A_i\right) \text{ iff it is false that } x \in \text{ at least one } A_i$$
$$\text{iff for all } i, x \notin A_i; \text{ i.e., } x \in A_i^c$$
$$\text{iff } x \in \bigcap_i A_i^c.$$

2. If $x \in A \cup (B \cap C)$, then either $x \in A$ or $x \in B \cap C$. In either case, $x \in (A \cup B) \cap (A \cup C)$. Conversely, let $x \in (A \cup B) \cap (A \cup C)$. If $x \in A$, then $x \in A \cup (B \cap C)$, so assume $x \notin A$. Then $x \in B$ and $x \in C$, so $x \in A \cup (B \cap C)$.

3. Amounts to a special case of Problem 4.

4. If $x \in \bigcup_{n=1}^{\infty} A_n$, then $x \in$ at least one A_i. If n is the smallest index such that $x \in A_n$, then $x \in A_1^c \cap \cdots \cap A_{n-1}^c \cap A_n$. Conversely, if x belongs to one of the sets $A_1^c \cap \cdots \cap A_{n-1}^c \cap A_n$, then, in particular, $x \in A_n$.

5. $$(A \cap B) \cup (A \cap C) \cup (A \cap D) \cup (B \cap C) \cup (B \cap D) \cup (C \cap D).$$

6. $$\begin{aligned}&(A \cap B \cap C^c \cap D^c) \cup (A \cap B^c \cap C \cap D^c) \cup (A \cap B^c \cap C^c \cap D)\\ &\cup (A^c \cap B \cap C \cap D^c) \cup (A^c \cap B \cap C^c \cap D) \cup (A^c \cap B^c \cap C \cap D).\end{aligned}$$

SECTION 1.2

1. Form S by specifying that $n \in S$ iff $n \notin S_n$, $n = 1, 2, \ldots$. If S is on the list, then $S =$ some S_{n_0}. Then $n_0 \in S$ iff $n_0 \in S_{n_0}$; but by definition of S, $n_0 \in S$ iff $n_0 \notin S_{n_0}$, a contradiction.

2. For each n, there are 2^n subsets of $\{1, 2, \ldots, n\}$ by Theorem 1.2.1. List them as $S_{n1}, S_{n2}, \ldots, S_{n2^n}$. Do this for each n, and enumerate all finite subsets as in Theorem 1.2.3.

3. Look at the diagonal where $x + y = 3$ (positions 6, 7, 8, 9 in the diagram). $\underbrace{\frac{1}{2}(x+y)(x+y+1)}_{6}$ is the sum of the first $x + y$ $(= 3)$ integers, which accounts for all previous diagonals ($x + y = 0$, 1, 2). Then x $(= 2x/2)$ locates the position within the diagonal; e.g., $x = 0$ yields position 6, $x = 1$ position 7, $x = 2$ position 8, $x = 3$ position 9. To go backwards, say we are given the integer 11. Since $1+2+3+4 = 10 \leq 11 < 1+2+3+4+5$, we are on the diagonal with $x + y = 4$; $x = 0$ gives position 10, $x = 1$ gives 11. Therefore $x = 1, y = 4 - 1 = 3$.

4. If the positive rationals are listed as $a_1, a_2, \ldots$, then all rationals will be given by $0, a_1, -a_1, a_2, -a_2, \ldots$.

5. There is no guarantee that the number r will be rational.

6. Consider two consecutive entries x_n and x_{n+1}. Then $r = \frac{1}{2}(x_n + x_{n+1})$ is a rational number with $x_n < r < x_{n+1}$, a contradiction.

7. $n = 1$: The rationals are countable.

   $n = 2$: Pairs of rationals are countable, by the same diagonal process that we used to count the rationals.

   $n = 3$: The triple (a, b, c) can be regarded as a pair $((a, b), c)$, so triples of rationals are countable.

Etc. (We are actually using *mathematical induction*, to be considered formally in Section 1.4.)

SECTION 1.3

1. You may use either the definitions or the intuitive idea that open means "does not contain any of its boundary points."

   (a) closed

   (b) open

   (c) neither

   (d) $\{x \in R : \underbrace{(x-1)(x-2) < 0}_{1<x<2}\}$ open

   (e) closed (note that if $x_n = n$ then $x_n \to \infty$, but $\infty \notin R$)

2. (a) open

   (b) closed

   (c) closed

   (d) neither

(a)

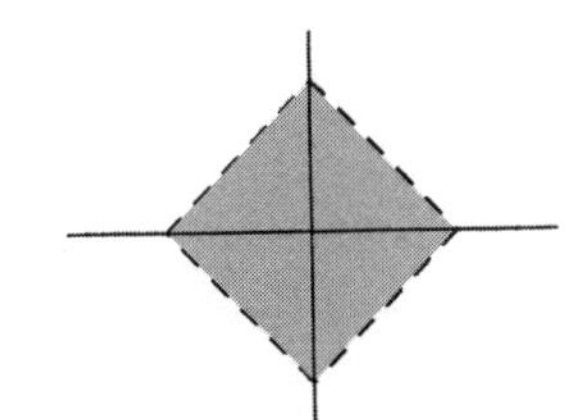

(b)

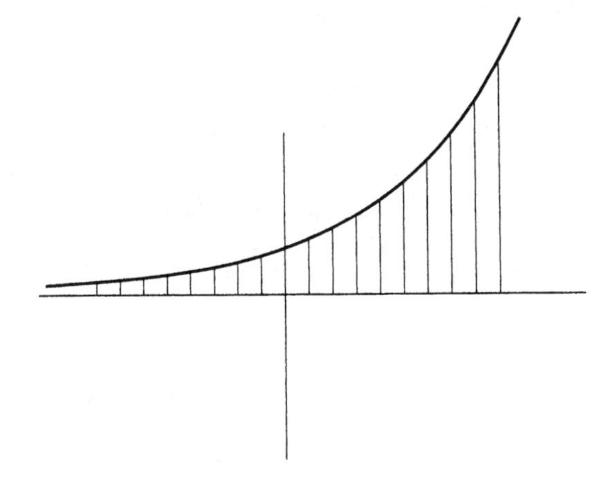

(c)

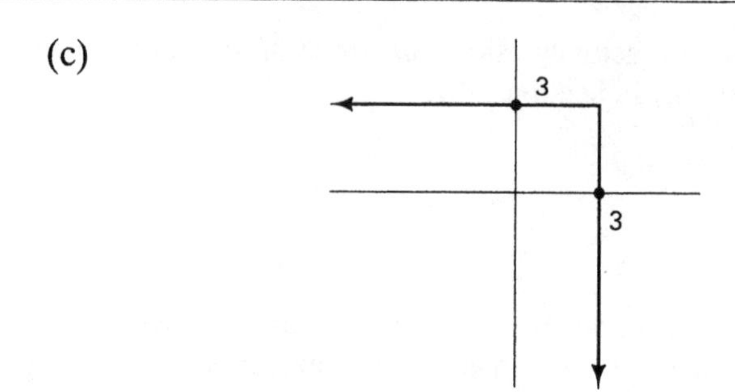

(d)

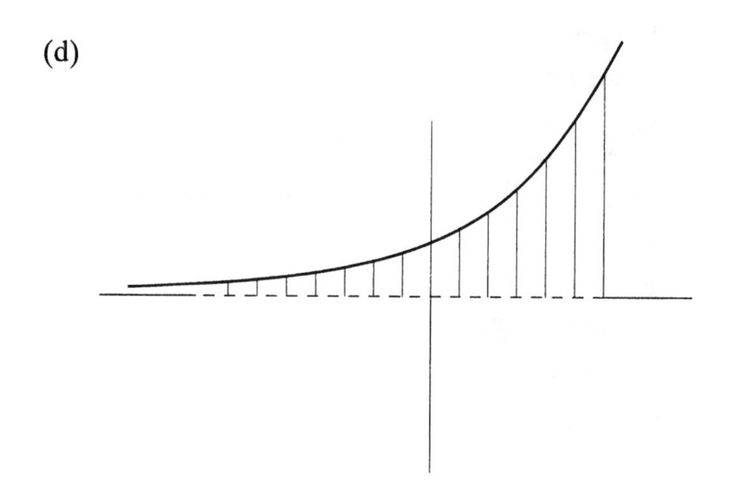

3. If $x \in R$, then any open interval $B_r(x)$ is $\subseteq R$, so R is open. If $x_n \in R$ and $x_n \to x \in R$, then $x \in R$ so R is closed.

   Note: $x = \infty$ is not legal here; ∞ is not a real number.

   The statements "R does not contain any of its boundary points" and "R contains every one of its boundary points" are both true since there are no boundary points.

4. The empty set $\emptyset$, by Theorem 1.3.2.

   Note: The statements "if $x \in \emptyset$, there is an open interval $B_r(x) \subseteq \emptyset$" and "if $x_n \in \emptyset$ and $x_n \to x \in R$, then $x \in \emptyset$" are *vacuously true* or *true by default*. For example, if you produce an $x \in \emptyset$, I will be happy to find an open interval $B_r(x) \subseteq \emptyset$. We will look at this idea more systematically in Section 1.4.

5. If $x \in E$, there is an open ball $B_r(x) = \{y \in \Omega : |y-x| < r\} \subseteq E\}$. (Note $B_r(1)$ = set of all y in Ω such that $|y-1| < r$, so for $0 < r < 1$, $B_r(1) = [1, 1+r)$.) If $x_n \in E$, $x_n \to x \in \Omega$, then $x \in E$, so E is closed in Ω (as well as closed in R).

6. Let $C = \{x_1, x_2, \ldots, x_n\}$. If $x \notin C$, say x_i is the point of C closest to x, with $r = d(x, x_i) > 0$. Then $y \in B_r(x)$ implies $y \notin C$, so $B_r(x) \subseteq C^c$. Thus C^c is open; hence C is closed.

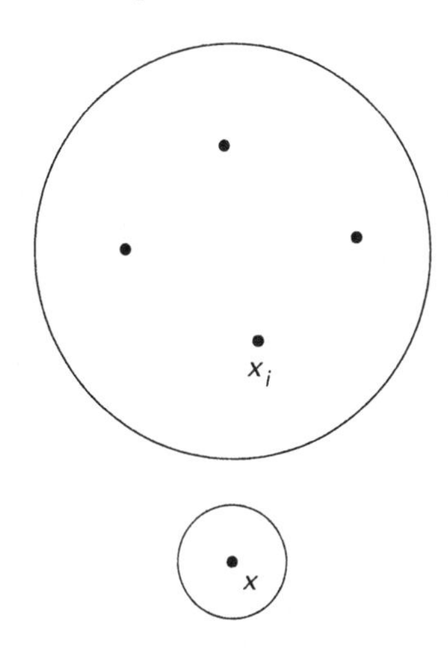

7. Assume $a \neq b$. If $B_{r_1}(a)$ and $B_{r_2}(b)$ are disjoint open balls, then $x_n \in$ both $B_{r_1}(a)$ and $B_{r_2}(b)$ for all sufficiently large n, which is impossible.

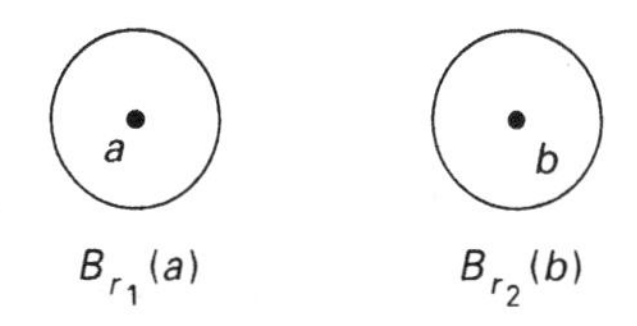

8. Fix $\epsilon > 0$. For sufficiently large n, a_n and b_n are within ϵ of L. Since $a_n \leq x_n \leq b_n$, it follows that $|x_n - L| < \epsilon$ for all sufficiently large n. Therefore $x_n \to L$.

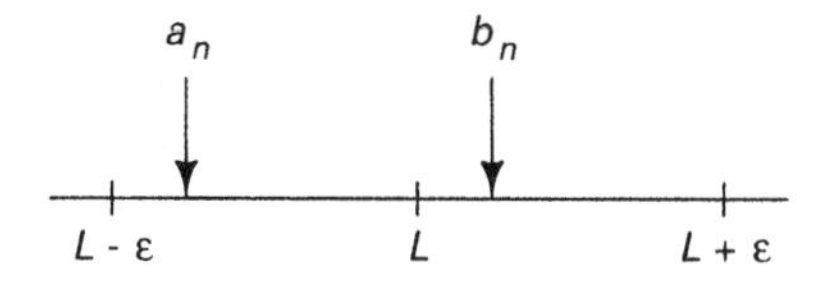

SECTION 1.4

1. There are many, many examples. A simple one is if $1 \leq x \leq 2$, then $1 \leq x \leq 3$.

2.

A	B	A and B	not (A and B)	not A	not B	not A or not B
T	T	T	F	F	F	F
T	F	F	T	F	T	T
F	T	F	T	T	F	T
F	F	F	T	T	T	T

A	B	A or B	not (A or B)	not A	not B	not A and not B
T	T	T	F	F	F	F
T	F	T	F	F	T	F
F	T	T	F	T	F	F
F	F	F	T	T	T	T

3.

P	Q	$P \to Q$	$Q \to P$	$P \leftrightarrow Q$
T	T	T	T	T
T	F	F	T	F
F	T	T	F	F
F	F	T	T	T

Thus, $P \leftrightarrow Q$ is T if and only if P and Q have the same truth value.

4. (a) True (for every positive real there is a bigger one).

   (b) False (there is no smallest positive real).

5. $\lim_{n\to\infty} d(x_n, x) = 0$ means

$$(\forall \epsilon > 0)(\exists N)(\forall n \geq N)(d(x_n, x) < \epsilon)$$

(here, ϵ is a (positive) real number, and N and n are positive integers).

6. $(\exists \epsilon > 0)(\forall N)(\exists n \geq N)(d(x_n, x) \geq \epsilon)$. There is a positive number ϵ such that for every positive integer N, $d(x_n, x) \geq \epsilon$ for some $n \geq$

N. In other words, no matter how far out we go in the sequence, we can find an element further along whose distance from x is at least ϵ. This expresses the idea that $d(x_n, x)$ does not approach 0.

7. *Case 1.* P holds.

   Case 2. Not P holds.

   In either case, we have "not P."

SECTION 1.5

1. Given any $B_r(x)$ we have $x_n \in B_r(x)$ for all sufficiently large n. Since $x_n \in E$, $x_n \neq x$, we can conclude that every open ball about x contains a point of E other than x, so $x \in E'$.
2. $E' = \bar{E} = \{(x,y) : 0 \leq x \leq 1, 0 \leq y \leq 1\}$.
3. $E' = \{0\}$, $\bar{E} = E \cup E' = \{1, \frac{1}{2}, \frac{1}{3}, \frac{1}{4}, \ldots, 0\}$.
4. If $x_1, x_2, \ldots \in E'$, $x_n \to x$, we must show $x \in E'$. Given $B_r(x)$, $x_n \in B_r(x)$ for all sufficiently large n. Pick any $x_n \in B_r(x)$ and let B be a ball about x_n entirely contained in $B_r(x)$ and with $x \notin B$. Since $x_n \in E'$, B contains a point y of E, and $y \neq x$ since $x \notin B$. Thus, $y \in B_r(x)$, $y \in E$, and $y \neq x$, so $x \in E'$.

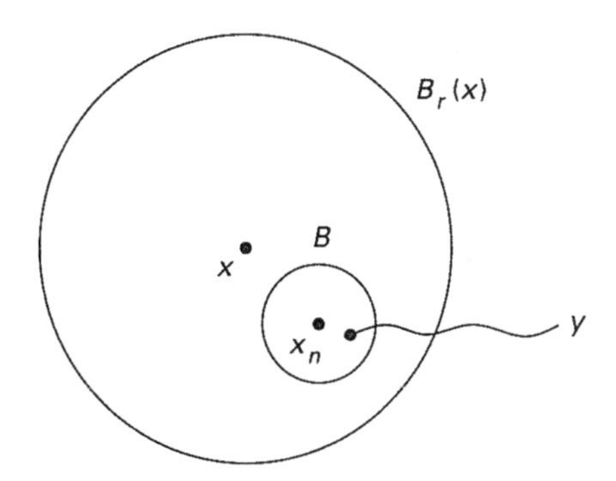

5. Assume x is a limit point of E. Every open ball $B_r(x)$ contains a point $y \in E$ with $y \neq x$. Since $E \subseteq \bar{E}$, we have $y \in \bar{E}$, so x is a limit point of $\bar{E}$. Now assume x is a limit point of $\bar{E}$. If $B_r(x)$ is an open ball about x, then $B_r(x)$ contains a point $y \in \bar{E}$ with $y \neq x$. Since $\bar{E} = E \cup E'$, we have $y \in E$ or $y \in E'$. If $y \in E$, we are finished, so assume $y \in E'$. Let B be a ball about y entirely contained in $B_r(x)$, with $x \notin B$. Then B will contain a point $z \in E$.

Thus, we have $z \in B_r(x)$, $z \in E$, $z \neq x$. We conclude that x is a limit point of E.

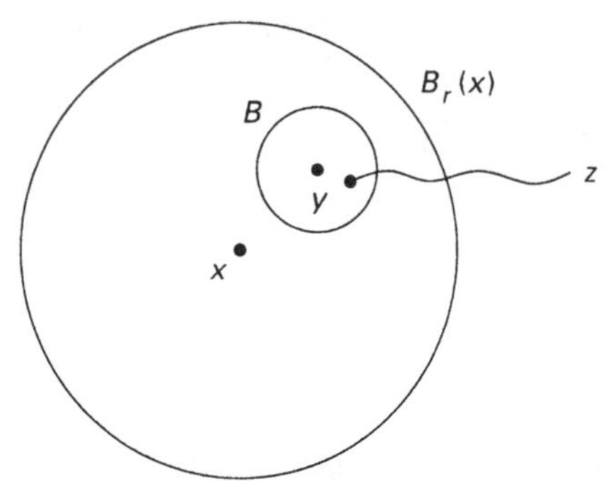

6. *Method 1.* Let x be a limit point of E'. If $B_r(x)$ is an open ball about x, then $B_r(x)$ contains a point $y \in E'$ with $y \neq x$. Exactly as in Problem 5 (last four sentences), x is a limit point of E.

   Method 2. By Theorem 1.5.1 (with E replaced by E') the set of limit points of E' is contained in the closure of E'. But by Problem 4, E' is closed, so its closure is E' itself. Thus, the set of limit points of E' is contained in E'; i.e., any limit point of E' belongs to E'.

7. (a) $E' = [0, 1] \cup \{2\}$.

   (b) 2 is a limit point of E but not a limit point of E'.

8. Let $x \in \bar{E}$, x not an interior point of E. Then $\exists x_n \in E$ with $x_n \to x$, so each $B_r(x)$ contains a point of E. Since x is not an interior point, $B_r(x) \not\subseteq E$ so $\exists y \in B_r(x)$ with $y \notin E$; i.e., $y \in E^c$.

   Now assume the condition "$x \in \bar{E}$, x not an interior point of E" is false.

   Case 1. $x \notin \bar{E}$. Then $x \notin E$ and $x \notin E'$, so some open ball $B_r(x)$ contains no point of E, and therefore x cannot be a boundary point.

   Case 2. x is an interior point of E. Then some $B_r(x) \subseteq E$, so $B_r(x)$ contains no point of E^c. Again, x cannot be a boundary point of E.

REVIEW PROBLEMS FOR CHAPTER 1

1. There is one–to–one correspondence between co–finite and finite subsets, so by Section 1.2, Problem 2, there are countably many co–finite subsets.

2. True. Any open ball $B_r(x)$ contains a point $y \in \bar{E}$ with $y \neq x$. If $y \in E$ we are finished, so assume $y \in E'$. But then, exactly as in Section 1.5, Problem 5, there is an open ball B about y entirely contained in $B_r(x)$, with $x \notin B$, and B contains a point $z \in E$. Since $x \notin B$, we have $z \neq x$, so $z \in B_r(x)$, $z \in E$, $z \neq x$, as desired.

3. Let $y \in B_r(x)$; we show that if s is sufficiently small then $B_s(y) \subseteq B_r(x)$. If $z \in B_s(y)$, then $d(z,y) < s$. By the triangle inequality,

$$d(z,x) \leq d(z,y) + d(y,x).$$

Now $d(y,x) < r$, so let $d(y,x) = r - \epsilon$, $\epsilon > 0$. Then $d(z,x) < s + r - \epsilon$, which is less than r if $s < \epsilon = r - d(x,y)$.

4. (a) $(\exists\epsilon > 0)(\forall N)(\exists n \geq N)(\exists m \geq N)(d(x_n,x_m) \geq \epsilon)$.

   (b) Take $N = 1$ to produce n_1, m_1 with $d(x_{n_1},x_{m_1}) \geq \epsilon$. Then take $N > \max(n_1,m_1)$ to produce n_2, m_2 with $d(x_{n_2},x_{m_2}) \geq \epsilon$. Continuing this process yields the desired sequence.

5. If E is a nonempty finite set, then all points of E are isolated.

SECTION 2.1

1. $\bigcup_{n=1}^{\infty}[0, 1 - 1/n] = [0,1)$.

2. If $x \in V$ where V is open, then there is an open ball $B(x)$ such that $x \in B(x)$ and $B(x) \subseteq V$. Since each point of V belongs to at least one of the balls, we have

$$V = \bigcup_{x \in V} B(x).$$

3. Suppose V_x and V_y are not disjoint (see diagram). If $z \in V_x \cap V_y$, then since $z \in V_x$, z belongs to some open interval I with $x \in I$ and $I \subseteq V$. Since $V_y \cap I$ is not empty (it contains z), $V_y \cup I$ is an open interval.

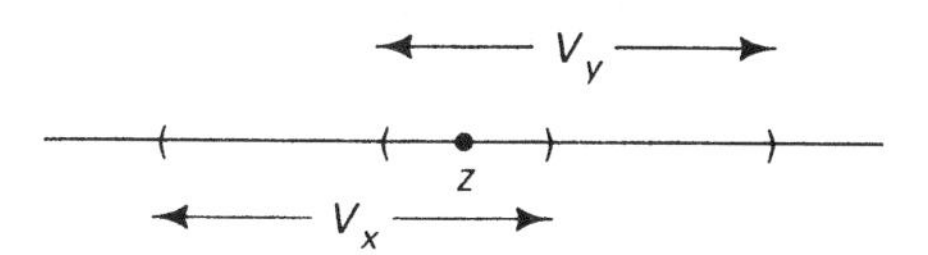

Since $x \in I$, we have $x \in V_y \cup I$, so $V_y \cup I$ is one of the open intervals involved in the definition of V_x, so $V_y \cup I \subseteq V$ and therefore

$V_y \subseteq V_x$. A symmetrical argument shows that $V_x \subseteq V_y$; hence $V_x = V_y$; i.e., V_x and V_y are identical.

4. The set S is formed by choosing a rational number from each V_x. Since the rationals are countable, the result follows.

SECTION 2.2

1. Only (c) is closed and bounded, hence compact ((a) and (e) are unbounded, (a), (b), and (d) are not closed).
2. Let $G_n = (1 - \delta, n)$, $n = 1, 2, \ldots$, where δ is any positive number.
3. (a) Let $G_n = \{(x,y) : x^2 + y^2 < 1 - 1/n\}$, $n = 1, 2, 3, \ldots$, so that the G_n form concentric circles whose radii approach 1. Any point in D belongs to G_n for all sufficiently large n.

   (b) Since D itself is open, we may take the covering to consist of D alone.
4. For compactness, *every* open covering must have a finite subcovering.
5. Yes. If $\emptyset \subseteq \bigcup_i G_i$, G_i open, then $\emptyset \subseteq$ any G_i.

SECTION 2.3

1. (a) 0

   (b) 0

   (c) $\pi/2$

   (d) e^{-3}

   (e) no limit
2. If $d(x,y) > k$, then $x \notin B_k(y)$. Choose r small enough so that $B_r(x) \cap B_k(y) = \emptyset$. Since $x_n \to x$, we have $x_n \in B_r(x)$ for all sufficiently large n. But then $d(x_n, y) > k$, a contradiction. Alternatively $d(x,y) \le d(x,x_n) + d(x_n,y)$ by the triangle inequality. Since $d(x,x_n) \to 0$ and $d(x_n,y) \le k$ for all n, the result follows.
3. Let $x_n = 1/n$, $n = 1, 2, \ldots$. (Every subsequence converges to $0 \notin (0,1)$.)
4. If ∞ belongs to the set A of the open covering, then for some r, $(r, \infty] \subseteq A$. Similarly, if $-\infty \in B$, then for some s, $[-\infty, s) \subseteq B$.

But $[s, r]$ is compact, so it can be covered by finitely many sets of the open covering. These, along with A and B, give a finite subcovering of $\bar{R}$.

5. For example, take $A = R$. By Problem 3 of Section 1.3, R is closed in R. But if $x_n = n$, then $x_n \in R$, $x_n \to \infty \notin R$, so R is not closed in $\bar{R}$.

6. No. Assume $A \subseteq R$ and A closed in $\bar{R}$. If $x_n \in A$, $x_n \to x \in R$, then $x \in \bar{R}$, so by assumption, $x \in A$.

SECTION 2.4

1. If b is a lower bound of E, then $-b$ is an upper bound of $-E$, and conversely. By the first part of the proof, $-E$ has a least upper bound $-c$. Since $-c \le -b$ for every upper bound $-b$ of $-E$, we have $b \le c$ for every lower bound b of E, so $c = \inf E$.

2. See the proof of Theorem 2.4.5 (here the convergent subsequence $\{x_{n_j}\}$ is given by hypothesis rather than the Bolzano–Weierstrass theorem).

3. By Theorem 2.3.1, any sequence in K has a convergent subsequence (with the limit in K). By Problem 2, any Cauchy sequence in K converges to a point of K.

4. Many examples, e.g., $\Omega_1 = \{x \in R : x \neq 0\}$; $\Omega_2 =$ the set of all rational numbers. Note that if $x_n = 1/n$, $n = 1, 2, \ldots$, then $\{x_n\}$ is a Cauchy sequence of points in Ω_1, but does not converge to a point *of* Ω_1.

5. Let $x_n \in K$, $x_n \to x$; show $x \in K$. By hypothesis there is a subsequence $x_{n_j} \to y \in K$. But since $x_n \to x$, we have $x_{n_j} \to x$; hence $x = y$. Thus $x \in K$, proving K closed. If K is unbounded, $\exists x_n \in K$ with $|x_n| \to \infty$. By hypothesis there is a subsequence $x_{n_j} \to y \in K$. But we also have $|x_{n_j}| \to \infty$, a contradiction since a convergent sequence is bounded.

   Note: A subsequence of $\{x_n\}$ is still a sequence: $y_1 = x_{n_1}, y_2 = x_{n_2}$, etc.

6. (a) Assume $x_n < 2$. Then $x_{n+1} = \sqrt{2 + \sqrt{x_n}} < \sqrt{2 + \sqrt{2}} < \sqrt{2 + 2} = 2$. Since $x_1 < 2$, we have, by induction, $x_n < 2$ for all n. Assume $x_n \ge x_{n-1}$. Then $x_{n+1} = \sqrt{2 + \sqrt{x_n}} \ge \sqrt{2 + \sqrt{x_{n-1}}} = x_n$. Since $x_2 = \sqrt{2 + \sqrt{x_1}} \ge \sqrt{2} = x_1$, $\{x_n\}$ is increasing, again by induction. The result now follows from Theorem 2.4.6.

(b) Let $n \to \infty$ in $x_{n+1} = \sqrt{2 + \sqrt{x_n}}$ to get $L = \sqrt{2 + \sqrt{L}}$, which can be solved numerically. (On a calculator, $L = 1.8311772$.)

7. If $a_x < \omega < b_x$, then by Theorem 2.4.3 there are numbers y, z such that $z < \omega < y$ and $(x,y) \subseteq V$, $(z,x) \subseteq V$ (hence, $(z,y) \subseteq V$, because x is assumed in V). But $\omega \in (z,y)$, so $\omega \in V_x$. Conversely, if $\omega \in V_x$, then for some open interval $I = (s,r)$, we have $\omega \in I$ and $x \in I \subseteq V$ (see diagram). Since $b_x \geq r$ and $a_x \leq s$, we have $a_x < \omega < b_x$.

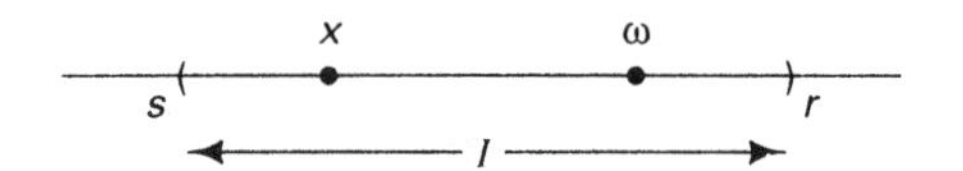

8. If $x \in V$ (open), then for some open interval I, $x \in I \subseteq V$; hence $x \in V_x$. Thus $V = \bigcup_{x \in V} V_x$. But by Section 2.1, Problem 4, there are only countably many distinct V_x, and the result follows.
9. False; e.g., let E be a finite set.
10. False; e.g., let $E = [0,1) \cup (2,3]$.

REVIEW PROBLEMS FOR CHAPTER 2

1. $[1, \infty)$ is closed and unbounded, hence not compact.
2. Let $I = (x - r, x + r)$ be an open interval about x. Since x is an upper bound, every $y \in E$ is less than or equal to x, hence less than x since $x \notin E$. If I contains no point of E, then $y \in E$ implies $y \leq x - r$, so $x - r$ is an upper bound of E, contradicting the fact that x is the least upper bound of E.
3. Let $x = \sup E$. If I is an open interval containing x, then I contains a point of E (if $x \notin E$, use Problem 2), and I also contains points of E^c (since any number greater than x is in E^c).
4. $L = 12/(1+L)$, so $L^2 + L - 12 = (L+4)(L-3) = 0$. By induction, $x_n > 0$ for all n, so L must be 3.
5. By induction, $1 < x_n < 2$ for all n; also,

$$x_{n+2} - x_{n+1} = 2 - \frac{1}{x_{n+1}} - (2 - \frac{1}{x_n}) = \frac{1}{x_n} - \frac{1}{x_{n+1}},$$

so $x_{n+1} < x_n$ implies $x_{n+2} < x_{n+1}$. Thus, $\{x_n\}$ is a bounded, decreasing sequence, so $x_n \to L$, where $L = 2 - 1/L$; i.e., $(L-1)^2 = 0$. Therefore $L = 1$.

6. For any x_0 we have $f(x_0) \le \sup_{x\in\Omega} f(x)$ and $g(x_0) \le \sup_{x\in\Omega} g(x)$, so

$$f(x_0) + g(x_0) \le \sup_{x\in\Omega} f(x) + \sup_{x\in\Omega} g(x) = c.$$

Thus, c is an upper bound of $A = \{f(x) + g(x) : x \in \Omega\}$, and therefore c is at least as big as the least upper bound of A, which is the desired result.

SECTION 3.1

1. $a_1, b_1, c_1, d_1, a_2, b_2, c_2, d_2, a_3, b_3, c_3, d_3, \ldots$, where $a_n \to -3$, $b_n \to 2$, $c_n \to 4$, $d_n \to 10$.

2. Say $f(n) \ge 2$ for $n \ge N$.

   Case 1. $x_N = 0$. Then $x_n = 0$ for all $n \ge N$, so $x_n \to 0$.

   Case 2. $x_N > 0$. Then $x_{N+1} \ge 2x_N, x_{N+2} \ge 4x_N, \ldots, x_{N+k} \ge 2^k x_N$, so $x_n \to \infty$.

   Case 3. $x_N < 0$. Then $x_{N+1} \le 2x_N, \ldots, x_{N+k} \le 2^k x_N$, so $x_n \to -\infty$. Thus, x_n always converges, and $\lim\sup x_n = \lim\inf x_n = 0$, ∞, or $-\infty$.

3. If S is the set of subsequential limits of $\{x_n\}$, the set of subsequential limits of $\{cx_n\}$ is $cS = \{cx : x \in S\}$. Thus, the largest subsequential limit of $\{cx_n\}$ is $c \lim\sup x_n$, and the smallest subsequential limit of $\{cx_n\}$ is $c \lim\inf x_n$.

4. Proceed as in Problem 3, noting that since c is *negative*, the largest subsequential limit of $\{cx_n\}$; i.e., the largest element of the form $cx, x \in S$, is c times the *smallest* element of S, i.e., $c \lim\inf x_n$. Similarly, the smallest subsequential limit of $\{cx_n\}$ is $c \lim\sup x_n$.

SECTION 3.2

1. If $\lim\inf y_n < z$, then by Theorem 3.2.2(d), $y_n < z$ i.o.; hence, $x_n < z$ i.o. Therefore some subsequence of $\{x_n\}$ has a limit $\le z$, so $\lim\inf x_n \le z$. Since z is any number greater than $\lim\inf y_n$, we have $\lim\inf x_n \le \lim\inf y_n$. If $\lim\sup y_n < z$, then by Theorem

3.2.2(a), $y_n < z$ ev.; hence $x_n < z$ ev. Therefore, all subsequential limits of $\{x_n\}$ are $\leq z$, so lim sup $x_n \leq z$. As above, conclude that $\limsup x_n \leq \limsup y_n$.

2. If $z > \limsup x_n$, $w > \limsup y_n$, then by Theorem 3.2.2(a), $x_n < z$ and $y_n < w$ ev., so $x_n + y_n < z + w$ ev. Thus, $\limsup(x_n + y_n) \leq z + w$. (If, e.g., $\limsup y_n = \infty$, take $w = \infty$.) Since z and w are arbitrary, as long as $z > \limsup x_n$, $w > \limsup y_n$, we have $\limsup(x_n + y_n) \leq \limsup x_n + \limsup y_n$. (Here we let $z \to \limsup x_n$, $w \to \limsup y_n$, and since $+\infty$ and $-\infty$ do not both appear on the right-hand side, $z + w \to \limsup x_n + \limsup y_n$.) The lim inf case is done exactly as above, with all inequalities reversed.

3.
$$0 \leq \sum_{j=1}^{n} (a_j x + b_j)^2 = \left(\sum_{j=1}^{n} a_j^2\right) x^2 + 2\left(\sum_{j=1}^{n} a_j b_j\right) x + \sum_{j=1}^{n} b_j^2.$$

Therefore, the quadratic polynomial on the right has either no real roots or at worst a real repeated root: so "$b^2 - 4ac$" is ≤ 0; i.e.,

$$4\left(\sum_{j=1}^{n} a_j b_j\right)^2 - 4\sum_{j=1}^{n} a_j^2 \sum_{j=1}^{n} b_j^2 \leq 0, \qquad \text{as desired.}$$

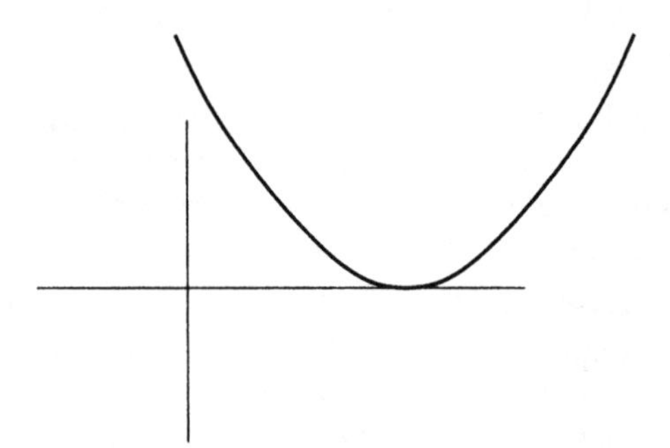

4.
$$\left[\sum_{n=1}^{\infty} \sqrt{a_n}\frac{1}{n}\right]^2 \leq \sum_{n=1}^{\infty} a_n \sum_{n=1}^{\infty} \frac{1}{n^2} \quad \text{by Problem 3}$$

(strictly speaking, apply Problem 3 to $\sum_{n=1}^{m}$, then let $m \to \infty$)

$< \infty$ by hypothesis.

Without Problem 3: If $\sqrt{a_n} \le 1/n$, then $\sqrt{a_n}/n \le 1/n^2$; if $\sqrt{a_n} > 1/n$, then $\sqrt{a_n}\sqrt{a_n} > \sqrt{a_n}(1/n) = \sqrt{a_n}/n$, so $\sqrt{a_n}/n < a_n$. Since $\sum 1/n^2 < \infty$ and $\sum a_n < \infty$, $\sum \sqrt{a_n}/n$ converges by comparison.

5.
$$|x+y|^2 = \sum_{i=1}^{n}(x_i+y_i)^2 = \sum_{i=1}^{n} x_i^2 + \sum_{i=1}^{n} y_i^2 + 2\sum_{i=1}^{n} x_i y_i.$$

By Problem 3,

$$|x+y|^2 \le |x|^2 + |y|^2 + 2|x||y| = (|x|+|y|)^2,$$

so $|x+y| \le |x| + |y|$.

Now

$$d(x,z) = |x-z| = |x-y+y-z| \le |x-y| + |y-z| = d(x,y) + d(y,z),$$

which proves the triangle inequality. The other properties of a metric are immediate.

6. "$x_n > 3$ eventually" can be expressed as

$$(\exists n)(\forall k \ge n)(x_k > 3).$$

The negation is

$$(\forall n)(\exists k \ge n)(x_k \le 3);$$

i.e., for every n, there is a $k \ge n$ with $x_k \le 3$, which says that $x_n \le 3$ infinitely often.

SECTION 3.3

1. Since $\left(\frac{1}{3}\right)^k / \left(\frac{1}{2}\right)^k = \left(\frac{2}{3}\right)^k \to 0$ and $(1/2)^{k+1}/(1/3)^k = \frac{1}{2}\left(\frac{3}{2}\right)^k \to \infty$, the limit involved in the ratio test does not exist. But

$$\left(\frac{1}{2^n}\right)^{1/(2n-1)} \to \frac{1}{\sqrt{2}} \quad \text{since}$$

$$\frac{1}{2n-1}\ln\frac{1}{2^n} = -\frac{n\ln 2}{2n-1} \to -\frac{1}{2}\ln 2 = \ln 2^{-1/2}$$

and

$$\left(\frac{1}{3^n}\right)^{1/2n} \to \frac{1}{\sqrt{3}} \quad \text{since } \frac{1}{2n}\ln\frac{1}{3^n} = -\frac{n\ln 3}{2n} \to -\frac{1}{2}\ln 3 = \ln 3^{-1/2}.$$

Thus, $a = \limsup |a_n|^{1/n} = 1/\sqrt{2} < 1$, and the root test gives convergence.

2. Use $1/r = \limsup |c_n|^{1/n}$.

   (a) $(n^k)^{1/n} \to 1$ since $(k \ln n)/n \to 0$, so $r = 1$.

   (b) $\left(3^n/n^5\right)^{1/n} \to 3$, so $r = \frac{1}{3}$.

3. $\limsup |c_n|^{1/n} \geq 3$, since at least one subsequential limit *is* ≥ 3. Thus $r \leq \frac{1}{3}$.

4.
$$\begin{aligned} S &= 1-\tfrac{1}{2}+\tfrac{1}{3}-\tfrac{1}{4}+\tfrac{1}{5}-\tfrac{1}{6}+\tfrac{1}{7}-\tfrac{1}{8}+\tfrac{1}{9}-\tfrac{1}{10}+\cdots \\ \tfrac{1}{2}S &= \quad\ \tfrac{1}{2} \quad\ -\tfrac{1}{4} \quad\ +\tfrac{1}{6} \quad\ -\tfrac{1}{8} \quad\ +\tfrac{1}{10}-\cdots \\ \hline \tfrac{3}{2}S &= 1 \quad +\tfrac{1}{3}-\tfrac{1}{2}+\tfrac{1}{5} \quad +\tfrac{1}{7}-\tfrac{1}{4}+\tfrac{1}{9} \quad\quad +\tfrac{1}{11}-\tfrac{1}{6}+\cdots \end{aligned}$$

5. Since the series converges but not absolutely,

$$\sum \text{positive terms} = \sum |\text{negative terms}| = \infty.$$

Given any real number r, add enough positive terms to get a partial sum $> r$, then add enough negative terms to drive the sum $< r$, then enough positive terms to get the sum $> r$, etc. Since the nth term $\to 0$, the "overshoot" or "undershoot" $\to 0$, so the rearranged series $\to r$. To achieve $+\infty$, add positive terms until the sum is > 1, then one negative term, then positive terms until the sum is > 2, then one negative term, etc. Divergence to $-\infty$ is achieved similarly. For oscillation, add positive terms until the sum is > 1, then negative terms until the sum is < -1, then positive terms until the sum is > 1, etc.

6. (a)

$$\sum_{n=0}^{m} c_n x^n = \sum_{n=0}^{m} (s_n - s_{n-1}) x^n = (1-x) \sum_{n=0}^{m-1} s_n x^n + s_m x^m.$$

Let $m \to \infty$ to obtain the desired result.

(b) $|f(x)-s| = |(1-x)\sum_{n=0}^{\infty}(s_n - s)x^n|$ by part (a) and the identity $(1-x)\sum_{n=0}^{\infty}x^n = 1$, $|x|<1$. Thus,

$$|f(x)-s| \le (1-x)\sum_{n=0}^{N}|s_n - s||x|^n + \frac{\epsilon}{2}(1-x)\sum_{n>N}x^n$$
$$\le (1-x)\sum_{n=0}^{N}|s_n - s|x^n + \frac{\epsilon}{2}, \qquad \text{as desired.}$$

(c) Without loss of generality assume $r = 1$ (consider $g(y) = \sum_{n=0}^{\infty}c_n(ry)^n$, $-1<y<1$). Given $\epsilon>0$, choose N as in part (b), and then choose $\delta>0$ so that

$$x > 1-\delta \Longrightarrow (1-x)\sum_{n=0}^{N}|s_n - s|x^n \ \left(\le (1-x)\sum_{n=0}^{N}|s_n - s|\right) < \frac{\epsilon}{2}.$$

Then $x > 1-\delta \Longrightarrow |f(x)-s| < \epsilon$.

REVIEW PROBLEMS FOR CHAPTER 3

1. (a) $1, 0, 1, 0, 1, 0, \ldots$
   (b) $I = (\frac{1}{2}, \frac{3}{2})$, $\{x_n\} = 1, 0, 1, 0, 1, 0, \ldots$.
2. $\sum_n (\frac{1}{2})^n z^n$.
3. False; e.g., $\sum_n n^n z^n$.
4. (a) False; e.g., $5, 0, 5, 0, 5, 0, \ldots$.
   (b) True (Theorem 3.2.2(b)).
   (c) True (Theorem 3.2.2(c)).
   (d) False; e.g., $x_n = 4 + 1/n$.
5. $1, -1, 2, -2, 3, -3, \ldots$.
6. $A = [-1,0] \cup [0,1] = [-1,1]$.
   $B = [-1,0] \cap [0,1] = \{0\}$.

SECTION 4.1

1. (a) $|x_n + y_n - (x + y)| \le |x_n - x| + |y_n - y| \to 0.$

   (b) $|x_n - y_n - (x - y)| \le |x_n - x| + |y - y_n| \to 0.$

   (c) $|x_n y_n - xy| = |x_n y_n - x y_n + x y_n - xy|$

   $= |(x_n - x) y_n + x(y_n - y)| \le |x_n - x||y_n| + |x||y_n - y|.$

   Since a convergent sequence is bounded, for some $M > 0$ we have $|y_n| \le M$ for all n; hence, the right-hand side $\to 0$.

   (d) This follows from (c) if we prove that $1/y_n \to 1/y$. Now

   $$\left|\frac{1}{y_n} - \frac{1}{y}\right| = \frac{|y - y_n|}{|y||y_n|},$$

   and since $y_n \to y \neq 0$ we have $|y_n| \ge \frac{1}{2}|y|$ for large enough n. Thus,

   $$\frac{|y - y_n|}{|y||y_n|} \le \frac{2}{|y|^2}|y - y_n| \to 0.$$

2. This is immediate from Problem 1. For example, if $x_n \to x$ then, by continuity, $f(x_n) \to f(x)$, $g(x_n) \to g(x)$. By Problem 1, $f(x_n) + g(x_n) \to f(x) + g(x)$, so $f + g$ is continuous at x. The difference, product, and quotient are handled similarly.

3. A polynomial is built from constant functions and the identity function ($I(x) = x$) using sums and products. By Problem 2, it suffices to show that if $h(x) = c$ and $I(x) = x$, then h and I are continuous. But if $x_n \to x$, then $h(x_n) = c \to c = h(x)$ and $I(x_n) = x_n \to x = I(x)$, as desired.

4. The statement is not true in general. If $y \in f(\bigcap_i A_i)$, then $y = f(x)$ for some $x \in \bigcap_i A_i$, so $y \in f(A_i)$ for every i, and therefore $f(\bigcap_i A_i) \subseteq \bigcap_i f(A_i)$. But the inclusion may be proper. For example, let $\Omega = \{1, 2, 3\}$, $\Omega' = \{1, 2\}$. If $f(1) = 1$, $f(2) = 2$, $f(3) = 1$, $A_1 = \{1, 2\}$, $A_2 = \{2, 3\}$, then

   $$f(A_1 \cap A_2) = f(\{2\}) = \{2\},$$
   $$f(A_1) \cap f(A_2) = \{1, 2\} \cap \{1, 2\} = \{1, 2\}.$$

   Note that $f(A \cap B) = f(A) \cap f(B)$ if f is one–to–one (if $z = f(x) = f(y)$, $x \in A$, $y \in B$, then $x = y$).

5. If $f: \Omega \to \Omega'$ is continuous on Ω and A is a closed subset of Ω', then $f^{-1}(A)$ is closed (see Theorem 4.1.6). In the given case, $A = \{c\}$.

SECTION 4.2

1. Yes, with

$$f(0) = \lim_{x\to 0} x \ \ln x = \lim_{x\to 0} \frac{\ln x}{1/x} = \lim_{x\to 0} \frac{1/x}{-1/x^2} = 0.$$

2. By Theorem 4.2.5, f can be extended to a (uniformly) continuous function on $\bar{E}$. Since $\bar{E}$ is closed and bounded, it is compact. By Theorem 4.2.1, $f(\bar{E})$ is compact and therefore closed and bounded. But $f(E) \subseteq f(\bar{E})$, so $f(E)$ is bounded.

3. Given $\epsilon > 0$ there exists $\delta > 0$ such that if $x, y \in E$ and $|x - y| < \delta$, then $|f(x) - f(y)| < \epsilon$. Since E is bounded (so that $\bar{E}$ is compact), it can be covered by finitely many (say M) open balls of diameter $< \delta$. But then $f(E)$ is covered by M open balls of diameter $< \epsilon$.

4. $x_n \to x$ means $(\forall\epsilon > 0)(\exists N)(\forall n \geq N)x_n \in B_\epsilon(x)$, so $x_n \not\to x$ means $(\exists\epsilon > 0)(\forall N)(\exists n \geq N)x_n \notin B_\epsilon(x)$. Thus, for some $\epsilon > 0$, $x_n \in B_\epsilon^c(x)$ for infinitely many n. By Theorem 2.3.1, $\{x_n\}$ has a subsequence converging to a limit $y \in K$. Since $d(x_n, x) \geq \epsilon$ on the entire subsequence, we have $d(y,x) \geq \epsilon$, so $y \neq x$. (If $d(y,x) < \epsilon$, then $d(x_n,x) \leq d(x_n,y) + d(y,x) < \epsilon$ for large n, a contradiction.)

5. Let $f(x_n) \to f(x)$. By Problem 4, if $x_n \not\to x$ there is a subsequence $x_{n_i} \to y \in K$ with $y \neq x$. But by continuity, $f(x_{n_i}) \to f(y)$; hence $f(x) = f(y)$. Since f is one-to-one, $x = y$, a contradiction. Conclude that $x_n \to x$.

6. Take $A = R$, $f(x) = e^x$; then $f(A) = (0, \infty)$, which is not closed.

SECTION 4.3

1. Look at a picture of the values of f (see diagram). f is increasing, but is discontinuous on the entire positive x and y axes.

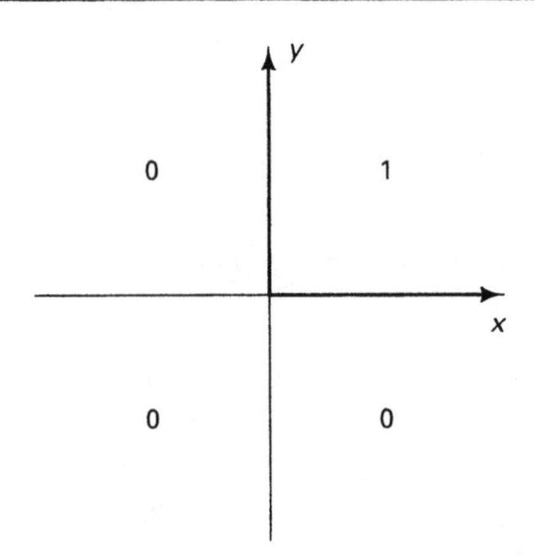

2. The only possible discontinuity is at $x = 0$. Now $|x \sin(1/x)| = |x||\sin(1/x)| \le |x| \to 0$ as $x \to 0$. Thus, $f(0^+) = f(0^-) = 0$, and $f(0) = 1$. Therefore, f has a removable discontinuity at 0.

3. Proceed as in Theorem 4.1.2. If the $\epsilon - \delta$ condition fails, there is an $\epsilon > 0$ such that for any $\delta > 0$, there is a point t within distance δ of x, with $t \neq x$, such that $d(f(t), A) \ge \epsilon$. If we take $\delta = 1/n$, $n = 1, 2, \ldots$, and label the corresponding t as t_n, then $t_n \neq x$, $t_n \to x$, but $f(t_n) \not\to A$. Now if the $\epsilon - \delta$ condition holds, let $t_n \neq x$, $t_n \to x$. Given $\epsilon > 0$, choose $\delta > 0$ so that $t \neq x$, $d(t,x) < \delta$ implies $d(f(t), A) < \epsilon$. Since $t_n \to x$, we have $d(t_n, x) < \delta$ ev. so $d(f(t_n), A) < \epsilon$ ev. It follows that $f(t_n) \to A$.

4. Assume f continuous at x. Given $\epsilon > 0, \exists \delta > 0$ such that $d(x,y) < \delta \Rightarrow d(f(x), f(y)) < \epsilon$. In particular, $x < y < x + \delta$ or $x - \delta < y < x \Rightarrow |f(x) - f(y)| < \epsilon$. Thus, $f(x^+) = f(x^-) = f(x)$. Conversely, if $f(x^+) = f(x^-) = f(x)$, then given $\epsilon > 0, \exists \delta_1, \delta_2 > 0$ such that

$$x < y < x + \delta_1 \Rightarrow |f(x) - f(y)| < \epsilon,$$
$$x - \delta_2 < y < x \Rightarrow |f(x) - f(y)| < \epsilon.$$

Take $\delta = \min(\delta_1, \delta_2) > 0$ and note that when $y = x$, $|f(x) - f(y)| = 0 < \epsilon$. Then $|x - y| < \delta \Rightarrow |f(x) - f(y)| < \epsilon$, so f is continuous at x.

SECTION 4.4

1. If $A = \{y\}$, then $d(x,A)$ is simply $d(x,y)$. The result follows from the discussion in Section 4.4.2.
2. At step n of the Cantor construction, the set *removed* consists of the union of 2^{n-1} intervals, each of length $1/3^n$ (so the length of the set removed is $\frac{1}{3}\left(\frac{2}{3}\right)^{n-1}$, and $\sum_{n=1}^{\infty}\frac{1}{3}\left(\frac{2}{3}\right)^{n-1} = 1$). We may modify the construction so that given α, $0 < \alpha < 1$, we remove the union of 2^{n-1} intervals, each of length $\frac{1}{2}\alpha\left(\frac{1}{4}\right)^{n-1}$. Then the length of the set removed is $\alpha\left(\frac{1}{2}\right)^n$, and $\sum_{n=1}^{\infty}\alpha\left(\frac{1}{2}\right)^n = \alpha$. The resulting Cantor-like set has length $1-\alpha$, and it can be shown that all properties in Theorem 4.4.1 (except (b)) still hold. (At Step 1, remove one interval of length $\alpha/2$; length $1-\alpha/2$ remains. Step 2: remove two intervals, each of length $\alpha/8$; OK since $\alpha/4 < 1-\alpha/2$; length $1-\alpha/2-\alpha/4$ remains. Step 3: remove four intervals, each of length $\alpha/32$; OK since $\alpha/8 < 1-\alpha/2-\alpha/4$; length $1-\alpha/2-\alpha/4-\alpha/8$ remains; etc.)
3. Choose $y_1, y_2, \ldots \in A$ with $d(x,y_n) \to d(x,A)$. By compactness, there is a subsequence $\{y_{n_k}\}$ converging to a limit $y_0 \in A$, so by continuity of distance, $d(x,y_{n_k}) \to d(x,y_0)$. But since $d(x,y_n) \to d(x,A)$, we have $d(x,y_{n_k}) \to d(x,A)$ also. By uniqueness of limits, $d(x,y_0) = d(x,A)$.
4. Let y be any point of A. Then $y \in$ some closed ball $C_r(x)$, and $d(x,A) = d(x, A\cap C_r(x))$, since points of A outside of $C_r(x)$ will be further away from x than y. Since $A \cap C_r(x)$ is compact, the result follows from Problem 3.

REVIEW PROBLEMS FOR CHAPTER 4

1. $f(x) = \sin\frac{1}{x-2}, x \neq 2; f(2) = 0.$
2. $x = 0, \quad A = (0,\infty).$
3. f is continuous at $x = 0$ (and discontinuous everywhere else). For if $x_n \to 0$, then $|f(x_n)| \le |x_n|$, so $f(x_n) \to 0$.
4. By Theorem 4.2.5, f has a continuous extension to the compact set $[0,7]$. The result follows from Corollary 4.2.2.

5. This follows from the Intermediate Value Theorem, since $f(-1) < 0$ and $f(0) > 0$.
6. No. If so, f would have an extension to a continuous real-valued function g on $[0, 1]$. But $\ln x \to -\infty$ as $x \to 0+$, so $g(0) = -\infty$, contradicting the fact that g is real-valued.
7. Let $f(x) = e^{x^2}$. If V is the complement of the set of integers, then $A = f^{-1}(V)$. Since V is open and f is continuous, A is open.
8. $\{1, \frac{1}{2}, \frac{1}{3}, \frac{1}{4}, \ldots\}$.
9. Take $C = \{0\}$; then $f^{-1}(C) = (-\infty, 0)$, which is not closed.
10. There are infinite discontinuities at $x = 0$ and $x = 1$ (see diagram).

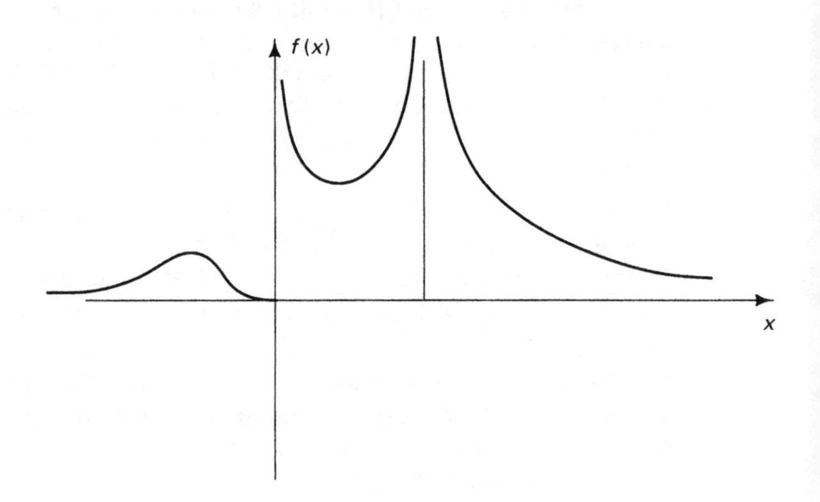

SECTION 5.1

1. The only possible difficulty is at $x = 0$. But

$$\frac{f(h) - f(0)}{h} = \frac{h^2 \sin(1/h)}{h} = h \sin\frac{1}{h} \to 0 \qquad \text{as} \quad h \to 0,$$

since $|h \sin(1/h)| = |h||\sin(1/h)| \leq |h|$. Thus $f'(0) = 0$.

2.

$$\left|\frac{f(x+h) - f(x)}{h}\right| \leq \frac{|h|^{1+r}}{|h|} = |h|^r \to 0.$$

Thus, $f' = 0$, so f is constant.

3. If $p(x) = p(y) = 0$ and $x < y$, then by Rolle's Theorem, $p'(z) = 0$ for some $z \in (x, y)$. Thus, if p has n distinct real roots, then p' has $n - 1$ distinct real roots. Since the degree of p' is $(\deg\ p) - 1$, there are no other roots of p'.

   If p has a repeated root of order k at x_0, then write $p(x) = (x - x_0)^k q(x)$, where $q(x_0) \neq 0$. But then $p'(x) = (x - x_0)^k q'(x) + k(x - x_0)^{k-1} q(x)$, so p' has a root of order $k - 1$ at x_0. Thus, $k - 1$ repetitions of the root of p correspond to a root of p' of order $k - 1$. So in any case, if p has n real roots counting multiplicity, then p' has $n - 1$ real roots counting multiplicity.

4. Let $f(x) = x^3$, $g(x) = x^2$, on $[-1, 1 + \delta]$. Then

$$\frac{f(1+\delta) - f(-1)}{g(1+\delta) - g(-1)} = \frac{(1+\delta)^3 - (-1)^3}{(1+\delta)^2 - (-1)^2}$$

$$= \frac{2 + 3\delta + 3\delta^2 + \delta^3}{2\delta + \delta^2} \to \infty \qquad \text{as } \delta \to 0.$$

   But

$$\frac{f'(x)}{g'(x)} = \frac{3x^2}{2x} = \frac{3}{2}x, \qquad \text{and} \qquad \left|\frac{f'(x)}{g'(x)}\right| \leq \frac{3}{2}(1 + \delta).$$

   Thus, if $\delta > 0$ is sufficiently small, $[f(1+\delta) - f(-1)]/[g(1+\delta) - g(-1)]$ can never equal $f'(x)/g'(x)$.

SECTION 5.2

1. There are many examples; e.g., $g(x) = 1 - e^{-x}$ with $a = 0$, $b > 0$.
2. Say $f'(x_0^-) = f'(x_0^+) = f'(x_0) - \epsilon$, $\epsilon > 0$. Choose $\delta > 0$ such that $f'(t) \leq f'(x_0) - \epsilon/2$ for $0 < |t - x_0| < \delta$ (see diagram). By Theorem 5.2.1 there exists $y \in (x_0 - \delta, x_0 + \delta)$ such that $f'(y) = f'(x_0) - \epsilon/4$, a contradiction.

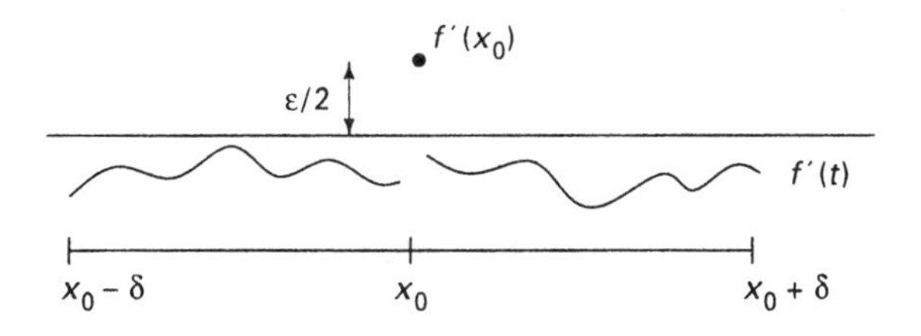

3. If $x \neq 0$, then

$$f'(x) = x^2 \left(\cos\frac{1}{x}\right)\left(-\frac{1}{x^2}\right) + 2x \sin\frac{1}{x}$$
$$= -\cos\frac{1}{x} + 2x \sin\frac{1}{x}; \qquad f'(0) = 0,$$

by Problem 1 of Section 5.1. Since $x \sin(1/x) \to 0$ as $x \to 0$, and $\cos(1/x)$ has a nonsimple discontinuity at $x = 0$ (look at the sequence $x_n = 1/n\pi$, for example), it follows that f' has a nonsimple discontinuity at $x = 0$.

4.

$$e^x = 1 + x + \frac{x^2}{2!} + \cdots + \frac{x^{n-1}}{(n-1)!} + R_n(x), \qquad R_n(x) = f^{(n)}(y)\frac{x^n}{n!}$$

for some y between 0 and x. But $f^{(n)}(y) = e^y$, so $|R_n(x)| \leq e^{|x|}x^n/n! \to 0$ as $n \to \infty$, so the series converges to e^x.

5. Let f be differentiable on R, with $|f'(x)| \leq M$ for all x. Fix $a \in R$. Then $f(b) - f(a) = (b - a)f'(x)$ for some x between a and b. Given $\epsilon > 0$, it follows that if $|b - a|M < \epsilon$, i.e., $|b - a| < \epsilon/M$, then

$$|f(b) - f(a)| < \epsilon.$$

REVIEW PROBLEMS FOR CHAPTER 5

1. No, although $f'(x) = 0$ for some $x \in (-1, 1)$, by Theorem 5.2.1. For an explicit counterexample, take $f(x) = x^2$.
2. The only difficulty occurs at $x = 0$. We have

$$\frac{f(h) - f(0)}{h} = \frac{e^{-1/h^2}}{h} \to 0 \qquad \text{as} \quad h \to 0$$

(let $h = 1/y$ and note that $ye^{-y^2} \to 0$ as $y \to \pm\infty$). Thus, $f'(x) = (2/x^3)e^{-1/x^2}, x \neq 0; f'(0) = 0$. A similar argument shows that

$$\frac{f'(h) - f'(0)}{h} \to 0 \qquad \text{as} \quad h \to 0,$$

and by repeating this procedure we obtain $f^{(n)}(0) = 0$ for all n. The key point is that e^{-1/h^2} will approach zero as $h \to 0$ faster than any polynomial can approach infinity.

3. See Problem 2.

4. Let $f(x) = 1 + 2x$, $g(x) = 1 + x$, or $f(x) = x + e^x$, $g(x) = e^x$. L'Hospital's rule does not apply since $f(x)$ and $g(x)$ do not approach 0 as $x \to 0$.

5. $f(x) = (x-1)^2 \sin \frac{1}{x-1}$, $x \neq 1$; $f(1) = 0$.

6. The remainder is of the form $f^{(n)}(y)x^n/n!$. Since $|f^{(n)}| \leq 1$ for all n and $x^n/n! \to 0$ as $n \to \infty$ for any fixed x, the result follows.

7. (a) We have $f(x)/g(x) = f'(y)/g'(z)$ and $f'(x)/g'(x) \to L$, but it does not follow (even if we ignore the problem of division by 0) that $f'(y)/g'(z) \to L$. For example, consider the two identical sequences

$$\{x_n\} = 1, \frac{1}{2}, \frac{1}{4}, \frac{1}{8}, \frac{1}{16}, \ldots,$$
$$\{y_n\} = 1, \frac{1}{2}, \frac{1}{4}, \frac{1}{8}, \frac{1}{16}, \ldots.$$

Then x_n/y_n is 1 for all n, but x_n/y_{n+1} is 2 for all n.

(b) If f' and g' are continuous at a, and $g'(a) \neq 0$, then

$$\frac{f(x)}{g(x)} = \frac{f'(y)}{g'(z)} \to \frac{f'(a)}{g'(a)} = L.$$

SECTION 6.1

1. Let

$$f(x) = \alpha(x) = \begin{cases} 1 & \text{if } x \geq 0, \\ 0 & \text{if } x < 0. \end{cases}$$

If $x_{k-1} \leq 0 \leq x_k$, then $M_k \, \Delta\alpha_k = 1$, $m_k \, \Delta\alpha_k = 0$, so if P is any partition of $[a, b]$ where $a < 0 < b$, then $U(P, f, \alpha) = 1$, $L(P, f, \alpha) = 0$. Thus, $\int_a^b f \, d\alpha$ does not exist.

2. Since any interval of positive length contains both rationals and irrationals, $U(P, f, \alpha)$ is always 1 and $L(P, f, \alpha)$ is always 0. Thus, $\int_0^1 f(x)\, dx$ does not exist.

3. Let $f(x) = 1, x$ rational; $f(x) = -1, x$ irrational. As in Problem 2, f is not Riemann integrable on $[0, 1]$, but $|f(x)| = 1$ for all x, so $|f|$ is Riemann integrable.

4. Given $\epsilon > 0$ and a partition P of $[a, b]$ with m subintervals, by definition of sup and inf (specifically by Theorem 2.4.3) we may choose $t_k^{(1)} \in [x_{k-1}, x_k]$ such that $(M_k - f_k)\ \Delta\alpha_k < \epsilon/2m$ and $t_k^{(2)} \in [x_{k-1}, x_k]$ such that $(f_k - m_k)\ \Delta\alpha_k < \epsilon/2m$. With the $t_k^{(1)}$, we get

$$(1) \qquad 0 \leq U(P, f, \alpha) - S(P^{(1)}, f, \alpha) < \epsilon/2,$$

and with the $t_K^{(2)}$, we get

$$(2) \qquad 0 \leq S(P^{(2)}, f, \alpha) - L(P, f, \alpha) < \epsilon/2.$$

But since $S(P, f, \alpha) \to \int_a^b f\ d\alpha$, it follows that $S(P^{(1)}, f, \alpha)$ and $S(P^{(2)}, f, \alpha)$ will differ by less than $\epsilon/2$ for sufficiently small $|P|$. But then

$$(3) \qquad 0 \leq U(P, f, \alpha) - L(P, f, \alpha) < \epsilon.$$

By (1), (2), and (3), U and L both $\to \int_0^b f\ d\alpha$.

SECTION 6.2

1. If $f \in R(\alpha)$, then $U(P), L(P) \to \int_a^b f\ d\alpha$; hence $U(P) - L(P) \to 0$. If $f \notin R(\alpha)$, then for some sequence of partitions P_n with $|P_n| \to 0$, we do not have $U(P_n), L(P_n) \to$ the same finite limit. By passing to convergent subsequences, we may assume that $U(P_n) \to U$, $L(P_n) \to L$, with $U \neq L$. But then $U(P_n) - L(P_n) \not\to 0$. (There is one other case to dispose of, namely, $U(P_n), L(P_n)$ both approach a limit, but for some other sequence $\{P_n'\}$, $U(P_n')$ and $L(P_n')$ both approach a different limit. A common refinement P_n'' of P_n and P_n' will decrease upper sums and increase lower sums (see the discussion before (2) of Section 6.1.1), which yields a contradiction. To spell this out, say $U(P_n)$ and $L(P_n) \to L_1$ and $U(P_n')$ and $L(P_n') \to L_2 > L_1$. Then $\{P_n''\}$ will have a subsequence on which $U(P_n'')$ will approach a limit $\leq L_1$, and $L(P_n'')$ will approach a limit $\geq L_2$, contradicting $L(P_n'') \leq U(P_n'')$.)

2. Since $U(P) - L(P)$ on $[a, c]$ and on $[c, b]$ are less than or equal to $U(P) - L(P)$ on $[a, b]$, the result follows from Problem 1.

3.

$$\int_0^3 f\,d\alpha = f(2)\frac{1}{3} + \int_2^3 f(x)\alpha'(x)\,dx$$
$$= \frac{4}{3} + \int_2^3 x^2 \frac{2}{3}\,dx = \frac{4}{3} + \frac{2}{3}\left(\frac{19}{3}\right) = \frac{50}{9}.$$

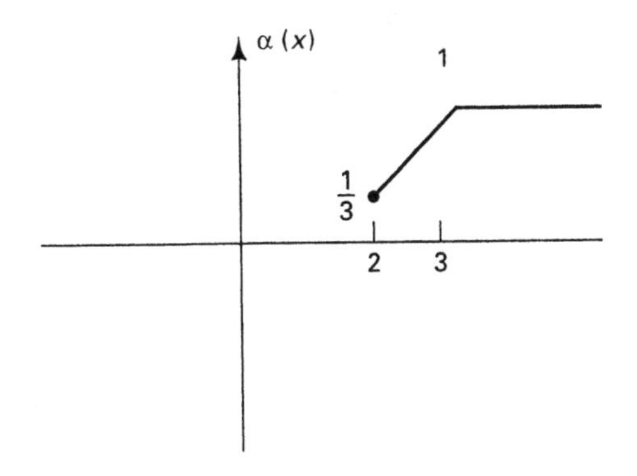

4. α has jumps of size 1 at all integers, so

$$\int_0^n x\,d\alpha(x) = 0 + 1 + 2 + \cdots + n = \frac{n(n+1)}{2}.$$

SECTION 6.3

1. By uniform continuity of f, there is a $\delta > 0$ such that $|x - y| < \delta$ implies $|f(x) - f(y)| < \epsilon/2$. By definition of variation, there is a partition P: $c = x_0 < x_1 < \cdots < x_n = b$ such that $\sum_{k=1}^{n} |f(x_k) - f(x_{k-1})| > V(f; [c, b]) - \epsilon/2$. Adding more points to P can only increase $\sum |f(x_k) - f(x_{k-1})|$, so we can assume $0 < x_1 - x_0 < \delta$; hence $|f(x_1) - f(x_0)| < \epsilon/2$.
2. $F(b) - F(c) = V(f; [c, b])$ by Theorem 6.3.2(e), so by Problem 1,

$$F(b) - F(c) - \frac{\epsilon}{2} < |f(x_1) - f(x_0)| + \sum_{k=2}^{n} |f(x_k) - f(x_{k-1})|$$
$$< \frac{\epsilon}{2} + V(f; [x_1, b]) = \frac{\epsilon}{2} + F(b) - F(x_1).$$

Therefore $F(x_1) - F(c) < \epsilon$. If $c < c' \leq x_1$, then, since F is increasing,

$$0 \leq F(c') - F(c) \leq F(x_1) - F(c) < \epsilon,$$

proving that F is right-continuous at c.

3. Choose δ as in Problem 1. Again by definition of variation, there is a partition P: $a = x_0 < x_1 < \cdots < x_n = c$ such that

$$\sum_{k=1}^{n} |f(x_k) - f(x_{k-1})| > V(f;[a,c]) - \frac{\epsilon}{2}.$$

As in Problem 1 we may add points to P so that $x_n - x_{n-1} < \delta$; hence $|f(x_n) - f(x_{n-1})| < \epsilon/2$.

4. $F(c) - F(a) = V(f;[a,c])$ by Theorem 6.3.2(e). By Problem 3,

$$F(c) - F(a) - \frac{\epsilon}{2} < |f(x_n) - f(x_{n-1})| + \sum_{k=1}^{n-1} |f(x_k) - f(x_{k-1})|$$
$$< \frac{\epsilon}{2} + V(f;[a,x_{n-1}]) = \frac{\epsilon}{2} + F(x_{n-1}) - F(a).$$

Thus, $F(c) < F(x_{n-1}) + \epsilon$. If $x_{n-1} \leq c' < c$, then $F(c) - F(c') \leq F(c) - F(x_{n-1})$, since F is increasing; hence $0 \leq F(c) - F(c') < \epsilon$, proving F left-continuous at c.

SECTION 6.4

1. By the Cauchy-Schwarz inequality for sums,

$$\left(\sum_{k=1}^{n} f(t_k)g(t_k)\sqrt{\Delta\alpha_k}\sqrt{\Delta\alpha_k}\right)^2 \leq \sum_{k=1}^{n} f^2(t_k)\,\Delta\alpha_k \sum_{k=1}^{n} g^2(t_k)\,\Delta\alpha_k.$$

Let $|P| \to 0$ to obtain the desired result. Alternatively,

$$\int_a^b (\lambda f + g)^2\, d\alpha = \left(\int_a^b f^2\, d\alpha\right)\lambda^2 + 2\left(\int_a^b fg\, d\alpha\right)\lambda + \int_a^b g^2\, d\alpha.$$

As in Section 3.2, Problem 3, the discriminant must be ≤ 0, and the result follows.

2. We have (see diagram)

$$f(n+1) \le \int_n^{n+1} f(x)\,dx \le f(n);$$

sum over n to get

$$\sum_{m=2}^{\infty} f(m) \le \int_1^{\infty} f(x)\,dx \le \sum_{n=1}^{\infty} f(n).$$

Since $\sum_{n=2}^{\infty} f(n) < \infty$ iff $\sum_{n=1}^{\infty} f(n) < \infty$, the series converges if and only if the integral is finite.

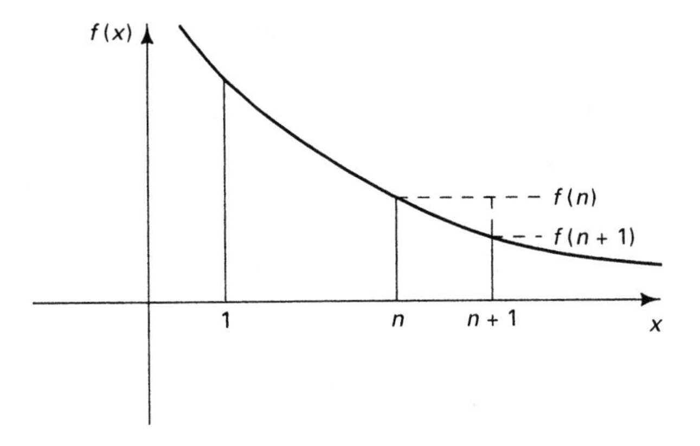

3. Let $f(x) = (-1)^{n-1}/n, \quad n-1 \le x < n, n = 1, 2, 3, \ldots$ (see diagram); we have $\int_0^{\infty} |f(x)|\,dx = \sum_{n=1}^{\infty} 1/n = \infty$, but $\int_0^{\infty} f(x)\,dx = 1 - \frac{1}{2} + \frac{1}{3} - \frac{1}{4} + \cdots$ (finite).

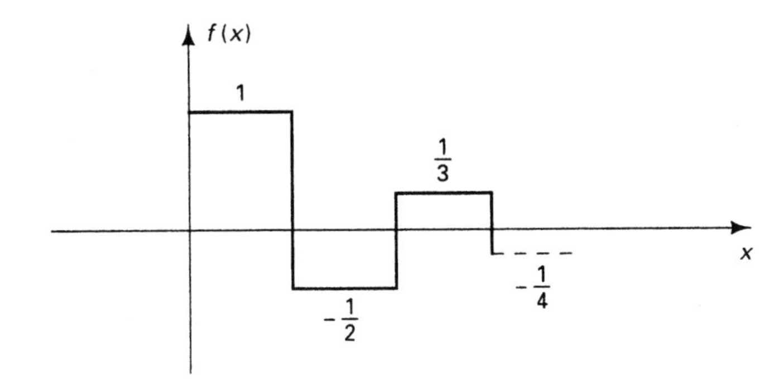

4. (a) Take $f(x) = 1/\sqrt{x}$.

   (b) Take $f(x) = 1/x$.

REVIEW PROBLEMS FOR CHAPTER 6

1.

$$\int_1^2 \frac{1}{x} 2x\,dx + \int_2^3 \frac{3}{x}\,dx + \int_3^4 \frac{1}{x} 3x^2\,dx$$
$$+ \frac{1}{2}[\alpha(2^+) - \alpha(2^-)] + \frac{1}{3}[\alpha(3^+) - \alpha(3^-)]$$
$$= 2 + 3(\ln 3 - \ln 2) + \frac{21}{2} + \frac{9-4}{2} + \frac{27-12}{3}$$
$$= 20 + 3(\ln 3 - \ln 2).$$

2. (a) False; f and α can be discontinuous at the same point.
   (b) True; see the discussion before (2) of Section 6.1.1.
   (c) True by Theorem 6.4.1, since $\int_a^b \alpha\,df$ exists.
3. Let F be an antiderivative of f; e.g., $F(x) = \int_a^x f(t)\,dt$. Then

$$\int_{g(x)}^{h(x)} f(t)\,dt = F(h(x)) - F(g(x)).$$

Differentiation yields, by the chain rule,

$$F'(h(x))h'(x) - F'(g(x))g'(x) = f(h(x))h'(x) - f(g(x))g'(x).$$

4. Let $f(x) = 1$ if x is rational; $f(x) = -1$ if x is irrational.
5.

$$\int_{-1}^0 3x(-2)\,dx + \int_0^3 3x(2)\,dx + \int_3^5 3x(0)\,dx + f(3)[\alpha(3+) - \alpha(3^-)]$$
$$= 3 + 27 + 9(10 - 6) = 66.$$

6. By the fundamental theorem of calculus, $F' = f$ on $[a, b]$. Therefore F has a bounded derivative on $[a, b]$, and the result follows from Theorem 6.3.2(c).
7. Our Change of Variable Theorem 6.4.2 breaks down when the substitution $y = h(x)$ is made and h is not one–to–one.

SECTION 7.1

1. $\lim_{n\to\infty}\lim_{x\to 0} f_n(x) = \lim_{n\to\infty} 0$ as before, but $\lim_{x\to 0}\lim_{n\to\infty} f_n(x) = \lim_{x\to 0} 1 = 1$.
2. $f_n \to f$ pointwise, where $f(x) = 0, 0 \le x < 1; f(1) = \frac{1}{5}$. Thus,

$$|f_n(x) - f(x)| = \begin{cases} \frac{x^n}{2+3x^n}, & 0 \le x < 1, \\ 0, & x = 1. \end{cases}$$

Since $|f_n(x) - f(x)| \to \frac{1}{5}$ as $x \to 1$, and $|f_n(x) - f(x)|$ increases with $x < 1$, $\sup_x |f_n(x) - f(x)| = \frac{1}{5}$. Thus $f_n \not\to f$ uniformly.
3. $\sup_{0\le x\le 1} |x^n/(n + x^n)| = 1/(n + 1) \to 0$, so $f_n \to 0$ uniformly.
4.

$$f_n(x) \to f(x) = \begin{cases} 1, & x > 1, \\ 0, & x = 1, \end{cases}$$

and

$$|f_n(x) - f(x)| = \begin{cases} \frac{1}{n+1}, & x = 1, \\ \frac{n}{n+x^n}, & x > 1. \end{cases}$$

Now $\sup_{x>1} n/(n + x^n) = n/(n + 1) \to 1$, so $f_n \not\to f$ uniformly on $[1, \infty)$. But $\sup_{x\ge 1+\delta} n/(n + x^n) = n/[n + (1 + \delta)^n] \to 0$, so $f_n \to f$ uniformly on $[1 + \delta, \infty)$.
5. $f_n(x) = (xe^{-x})^n$, and the maximum value of xe^{-x} occurs at $x = 1$. Thus, $\sup_{x\ge 0} f_n(x) = e^{-n} \to 0$, so $f_n \to 0$ uniformly on $[0, \infty)$.

SECTION 7.2

1. Take $f_n(x) = n$ for all x. Then $f_n'(x) \equiv 0$, so f_n' converges uniformly, although f_n does not.
2. Let $f_n \to f$, $g_n \to g$ uniformly. Then $|f_n + g_n - (f + g)| \le |f_n - f| + |g_n - g|$, so $f_n + g_n \to f + g$ uniformly.
3. Let $f_n(x) = 1/n \to 0$ uniformly on R; $g_n(x) = x \to x$ uniformly on R. But $f_n(x)g_n(x) = x/n \to 0$ pointwise but not uniformly on R, since $\sup_{x\in R} |x/n| = \infty$.
4. Say $f_n \to f$ uniformly. Fix $\epsilon > 0$; eventually, $|f_n - f| \le \epsilon$, say for $n \ge N$. Then $|f(x)| \le |f_N(x)| + \epsilon \le M_N + \epsilon < \infty$, so f is bounded. But then $|f_n| \le M_n$ for $n < N$, and $|f_n| \le |f| + \epsilon \le M_N + 2\epsilon$ for

$n \geq N$. Thus, $|f_n(x)| \leq \max(M_1, ..., M_{N-1}, M_N + 2\epsilon)$ for all n and all x.

5. Let $f_n(x) = x$ for all $n = 1, 2, \ldots$ and all $x \in R$; $f(x) = x$. Then $f_n \to f$ uniformly on R, but the f_n are not uniformly bounded.

SECTION 7.3

1. The Weierstrass M-test applies, since $e^{-nx} \leq e^{-na}$, and $\sum_n e^{-na} < \infty$ if $a > 0$.

2. The nth partial sum is

$$s_n(x) = \sum_{k=0}^{n} e^{-kx} = 1 + e^{-x} + \cdots + (e^{-x})^n = \frac{1 - (e^{-x})^{n+1}}{1 - e^{-x}}$$

$$\to s(x) = \frac{1}{1 - e^{-x}} \quad \text{as} \quad n \to \infty.$$

Thus, $|s_n(x) - s(x)| = e^{-(n+1)x}/(1 - e^{-x})$, and since $1 - e^{-x} \to 0$ as $x \to 0$, we have $\sup_{x>0} |s_n(x) - s(x)| = \infty$. Therefore, $s_n \not\to s$ uniformly on $(0, \infty)$.

3. This follows immediately from Theorem 3.3.2, since the radius of convergence is at least $|r|$.

4. On $[-a, a]$, $|a_n x^n| \leq |a_n| a^n$, and $\sum_{n=0}^{\infty} |a_n| a^n < \infty$ by Problem 3. The result follows from the Weierstrass M-test.

5. This follows from Theorems 7.2.2 and 7.2.3. To apply Theorem 7.2.3, we need only check that the differentiated series $\sum n a_n x^{n-1}$ has the same radius of convergence as the original series. But this follows from Theorem 3.3.2, since $\limsup |n a_n|^{1/n} = \limsup |a_n|^{1/n}$ (note that $n^{1/n} \to 1$).

6. $\int_a^b \left(\sum_{k=1}^{n} f_k\right) d\alpha = \sum_{k=1}^{n} \int_a^b f_k \, d\alpha$ by Theorem 6.2.1(a). Let $n \to \infty$; by Theorem 7.2.2, $\int_a^b \left(\sum_{k=1}^{\infty} f_k\right) d\alpha = \lim_{n\to\infty} \sum_{k=1}^{n} \int_a^b f_k \, d\alpha$; i.e., the series $\sum_{n=1}^{\infty} \int_a^b f_n \, d\alpha$ converges to $\int_a^b \left(\sum_{n=1}^{\infty} f_n\right) d\alpha$.

SECTION 7.4

1. Take $f_n(x_1) = 1$ for all n, $f_n(x_2) = 2$ for all n, $f_n(x_3) = 3$ for all n, etc.

2. (a) See diagram.

(b) $\{f_n\}$ converges pointwise on the dyadic rationals $k/2^n$, $k = 0, 1, \ldots, 2^n$, $n = 1, 2, \ldots$; $\{f_n\}$ is uniformly bounded since $0 \le f_n \le 1$ for all n. If the second and third digits of the binary expansion of x are both 1, then $f_1(x) \ge \frac{3}{4}$; if the digits are both 0, then $f_1(x) \le \frac{1}{4}$. Similarly, if the $(n+1)$st and $(n+2)$nd digits of the binary expansion of x are both 1, then $f_n(x) \ge \frac{3}{4}$; if the digits are both 0, then $f_n(x) \le \frac{1}{4}$. Thus, given any subsequence $\{f_{n_k}\}$, we may assume without loss of generality that $n_{j+1} > 2 + n_j$ for all j, and we can select the digits in the binary expansion of x to produce oscillation in the values of $f_{n_k}(x)$, so $\{f_{n_k}\}$ cannot converge pointwise.

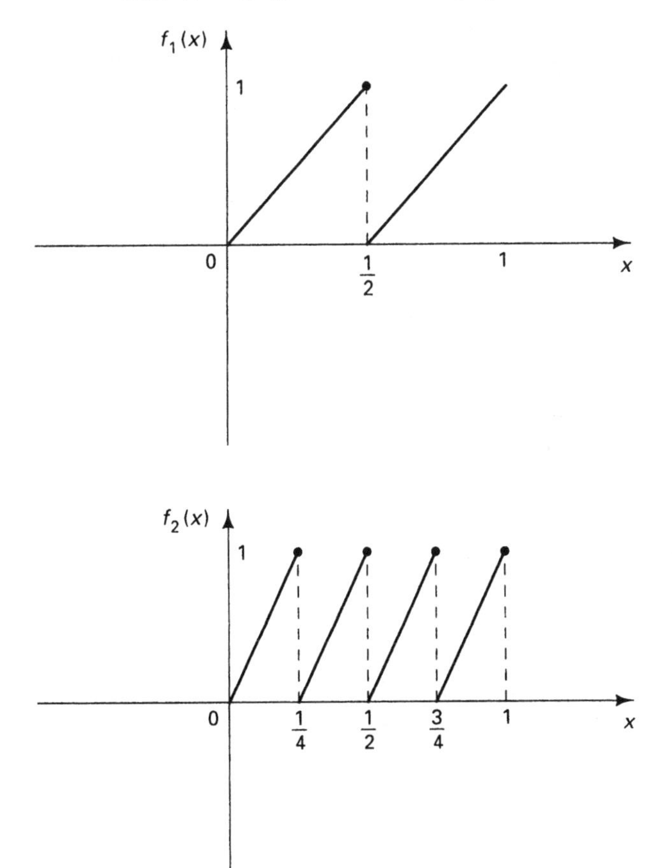

3. Given $\epsilon > 0$, choose $\delta > 0$ such that $d(x,y) < \delta$ implies $d(f_n(x), f_n(y)) < \epsilon/3$ for all n. If $x, y \in K$ and $d(x,y) < \delta$, then $d(f(x),$

$f(y)) \le d(f(x), f_n(x)) + d(f_n(x), f_n(y)) + d(f_n(y), f(y)) < \epsilon$ for sufficiently large n, by pointwise convergence. Thus f is uniformly continuous on K. Now adjust the choice of δ so that if $d(x,y) < \delta$, then, in addition to $d(f_n(x), f_n(y)) < \epsilon/3$ for all n, we have $d(f(x), f(y)) < \epsilon/3$. By compactness, there are points $x_1, \dots, x_m \in K$ such that $K \subseteq \bigcup_{i=1}^m B_\delta(x_i)$. Then

$$d(f(x), f_n(x)) \le d(f(x), f(x_i)) + d(f(x_i), f_n(x_i)) + d(f_n(x_i), f_n(x)).$$

If $x \in B_\delta(x_i)$, then $d(x, x_i) < \delta$, so the first and third terms on the right are $< \epsilon/3$. The second term is $< \epsilon/3$ for large n, say $n \ge N$, by pointwise convergence. It follows that $d(f(x), f_n(x)) < \epsilon$ for $n \ge N$ and all x, proving uniform convergence.

4. (a)

$$d(f_n(x), f_n(y)) \le d(f_n(x), f(x)) + d(f(x), f(y)) + d(f(y), f_n(y)).$$

Given $\epsilon > 0$, there is a positive integer N such that the first and third terms on the right are $< \epsilon/3$ for $n \ge N$ and all x, $y \in K$ (by uniform convergence). For some $\delta > 0$, $d(x,y) < \delta$ implies $d(f(x), f(y)) < \epsilon/3$, by uniform continuity of f. (f is a uniform limit of continuous functions, so is continuous. Since K is compact, f is uniformly continuous.) Since $f_1, \dots, f_{N-1}$ are uniformly continuous on K (compactness is also used here), there is a $\delta > 0$ such that $d(x,y) < \delta$ implies $d(f_n(x), f_n(y)) < \epsilon$ for all n, proving equicontinuity.

(b)

$$d(f_n(x_n), f(x)) \le d(f_n(x_n), f(x_n)) + d(f(x_n), f(x)).$$

Given $\epsilon > 0$, the first term on the right is $< \epsilon/2$ for large n, by uniform convergence. The second term is $< \epsilon/2$ for large n since f, a uniform limit of continuous functions, is continuous. Thus, $d(f_n(x_n), f(x)) < \epsilon$ for large n.

(c) The argument of (a) uses compactness, but (b) does not. For an explicit counterexample to (a), take $f_n(x) = f(x) = 1/x$ on $(0, \infty)$ for all n. Then $\{f_n\}$ is not equicontinuous because $x \to 1/x$ is not uniformly continuous on $(0, \infty)$. (See the discussion before Definition 4.2.3.)

SECTION 7.5

1. Let $g(y) = f(a + (b-a)y)$, $0 \le y \le 1$. If p is a polynomial such that $|g(y) - p(y)| < \epsilon$ on $[0,1]$, then $|f(x) - p\left[\frac{(x-a)}{(b-a)}\right]| =$ (with $x = a + (b-a)y$) $|g(y) - p(y)| < \epsilon$ for $a \le x \le b$.
2. One example: let $a_{nn} = n$, $a_{n,n+1} = -n$, $n = 1, 2, \ldots$; $a_{nj} = 0$ for all other pairs (n, j) of positive integers. Then

$$\sum_{n=1}^{\infty}\sum_{j=1}^{\infty} a_{nj} = \sum_{n=1}^{\infty}(a_{nn} + a_{n,n+1}) = 0;$$

$$\sum_{j=1}^{\infty}\sum_{n=1}^{\infty} a_{nj} = \sum_{j=1}^{\infty} 1 = \infty;$$

$$\begin{matrix} 1 & -1 & 0 & & & \cdots \\ 0 & 2 & -2 & 0 & & \cdots \\ 0 & 0 & 3 & -3 & & \cdots \\ 0 & 0 & 0 & 4 & -4 & \cdots \\ & \vdots & & & & \end{matrix}$$

3. For every $x \in E$, $|f_n(x)| \le M_n = \sum_{j=1}^{\infty} |a_{nj}|$. Since $\sum_{n=1}^{\infty} M_n < \infty$ by hypothesis, the result follows.
4. Since the x_k, $k \ge 1$, are isolated points, the f_n are automatically continuous there (if x_j converges to the isolated point y, then $x_j = y$ eventually). So all that needs to be verified is that $f_n(x_k) \to f_n(x_0)$ as $k \to \infty$. But, by hypothesis, $\sum_{j=1}^{\infty} |a_{nj}| < \infty$ for each n; in particular, $\sum_{j=1}^{\infty} a_{nj}$ converges. But then $f_n(x_k) = \sum_{j=1}^{k} a_{nj} \to \sum_{j=1}^{\infty} a_{nj} = f_n(x_0)$.
5. $\sum_{n=1}^{m}\sum_{j=1}^{k} a_{nj} = \sum_{j=1}^{k}\sum_{n=1}^{m} a_{nj}$ for any m. Since

$$\sum_{n=1}^{\infty}\left|\sum_{j=1}^{k} a_{nj}\right| \le \sum_{n=1}^{\infty}\sum_{j=1}^{\infty} |a_{nj}| < \infty,$$

and

$$\sum_{n=1}^{\infty} |a_{nj}| = \lim_{m\to\infty}\sum_{n=1}^{m} |a_{nj}| \le \sum_{n=1}^{\infty}\sum_{j=1}^{\infty} |a_{nj}| < \infty,$$

we may let $m \to \infty$ to get the desired result.

6.

$$f(x_k) = \sum_{n=1}^{\infty} f_n(x_k) = \sum_{n=1}^{\infty}\sum_{j=1}^{k} a_{nj},$$

$$f(x_0) = \sum_{n=1}^{\infty} f_n(x_0) = \sum_{n=1}^{\infty}\sum_{j=1}^{\infty} a_{nj}.$$

By Problem 4, $f(x_k) \to f(x_0)$ as $k \to \infty$, as desired.

7.

$$\sum_{n=1}^{\infty}\sum_{j=1}^{\infty} a_{nj} = \lim_{k\to\infty} \sum_{n=1}^{\infty}\sum_{j=1}^{k} a_{nj} \qquad \text{by Problem 6}$$

$$= \lim_{k\to\infty} \sum_{j=1}^{k}\sum_{n=1}^{\infty} a_{nj} \qquad \text{by Problem 5}$$

$$= \sum_{j=1}^{\infty}\sum_{n=1}^{\infty} a_{nj}.$$

The result is finite because all partial sums are bounded by $\sum_{n=1}^{\infty}\sum_{j=1}^{\infty} |a_{nj}| < \infty$.

8. Choose N and k so that s_{Nk} will be "close" to s. If s is finite, this means $|s - s_{Nk}| < \epsilon$; if $s = \infty$, take $s_{Nk} > M$, where M is an arbitrary positive number. Since $a_{nj} \geq 0$, s_{Nk} increases with N and k, and it follows that the double sum (in either order) is arbitrarily close to s.

9.

$$\sum_{n=0}^{\infty} |c_n||x|^n \leq \sum_{n=0}^{\infty}\sum_{k=0}^{n} |a_k||x|^k |b_{n-k}||x|^{n-k}$$

$$= \sum_{k=0}^{\infty} |a_k||x|^k \sum_{n=k}^{\infty} |b_{n-k}||x|^{n-k} \qquad \text{by Problem 8}$$

$$< \infty \qquad \text{for } |x| < r,$$

since the power series $\sum a_n x^n$ and $\sum b_n x^n$ converge absolutely with their interval of convergence. Now

$$h(x) = \sum_{n=0}^{\infty} c_n x^n = \sum_{n=0}^{\infty}\sum_{k=0}^{n} a_k x^k b_{n-k} x^{n-k}$$

$$= \sum_{k=0}^{\infty} a_k x^k \sum_{n=k}^{\infty} b_{n-k} x^{n-k}$$

$$= \sum_{k=0}^{\infty} a_k x^k \sum_{\ell=0}^{\infty} b_\ell x^\ell$$

$$= f(x)g(x), \qquad |x| < r.$$

REVIEW PROBLEMS FOR CHAPTER 7

1. Since $x^n \to 0$ for $0 \le x < 1$, $f_n(x) \to \frac{2}{3}$ pointwise. Now

$$\frac{2+x^n}{3+x^n} - \frac{2}{3} = \frac{x^n}{3(3+x^n)}, \qquad \text{and} \qquad \sup_{0 \le x < 1} \left| \frac{x^n}{3(3+x^n)} \right| = \frac{1}{12} \not\to 0$$

(note $x^n/(3+x^n) = 1 - 3/(3+x^n)$, which increases with x), so $\{f_n\}$ does not converge uniformly on $[0,1)$.

2. Take $f_n(x) = (-1)^n/n$ for all real x.

3. For any fixed x, $f_n(x)$ is eventually 0, so $f_n \to 0$ pointwise. But $\sup_{x \in R} f_n(x) = 1 \not\to 0$, so $f_n \not\to 0$ uniformly.

4. Power series converge uniformly on any closed subinterval of the interval of convergence.

SECTION 8.1

1. Let $x \in A \cap F$. Since A is open in E, for some $r > 0$ we have $\{y \in E : d(x,y) < r\} \subseteq A$. But $F \subseteq E$, so $\{y \in F : d(x,y) < r\} \subseteq A$ also. But then $\{y \in F : d(x,y) < r\} \subseteq A \cap F$, proving $A \cap F$ open in F.

2. Let $x_n \in A \cap F$, $x_n \to x \in F$. Then $x_n \in A$, $x_n \to x \in E$, so (because A is closed in E) $x \in A$. But we already have $x \in F$, so $x \in A \cap F$, proving $A \cap F$ closed in F.

3. If $h(x) = \arctan x$, then h is a homeomorphism of $\bar{R}$ and $[-\pi/2, \pi/2]$. Thus, $h \circ f$ is a bounded, continuous function on E, so by Theorem 8.1.3, there is a continuous function g_0 on Ω such that $g_0 = h \circ f$ on E and $\sup\{|g_0(x)| : x \in \Omega\} = \sup\{|h(f(x))| : x \in E\}$. If $g = h^{-1} \circ g_0$, then g is continuous on Ω and $g = f$ on E. (Note that if $g_0(x) = \pi/2$, then $g(x) = h^{-1}(g_0(x)) = \infty$, and similarly, if $g_0(x) = -\pi/2$, then $g(x) = -\infty$, so we cannot replace the codomain of g by R.)

SECTION 8.2

1. Let $y \in \overline{B_r(x)}$, so that we have points $y_n \in B_r(x)$ with $y_n \to y$. But then $d(x,y_n) \to d(x,y)$. (See Section 4.4, Problem 1.) Since $d(x,y_n) < r$ for all n, we have $d(x,y) \le r$.

2. Let Ω be any set with at least two points, and put the following metric on Ω:

$$d(x,y) = \begin{cases} 1, & x \neq y, \\ 0, & x = y. \end{cases}$$

Then $B_1(x) = \{x\} = \overline{B_1(x)} \subset \{y : d(x,y) \leq 1\} = \Omega$.

3. The result follows from the definition of convergence:

$$f_n(x) \to f(x) \quad \text{iff} \quad \forall m \exists n (\forall k \geq n)(|f_k(x) - f(x)| < \frac{1}{n}).$$

(Note that if for every positive integer m, $|f_n(x) - f(x)|$ is eventually less than $1/m$, then for every $\epsilon > 0$, $|f_n(x) - f(x)|$ is eventually less than ϵ.)

SECTION 8.3

1. Assume $\bar{A} \cap B = A \cap \bar{B} = \emptyset$. Since $A \subseteq \bar{A}$, $B \subseteq \bar{B}$, the sets A and B are disjoint. If $x_n \in A$, $x_n \to x \in A \cup B$, then $x \in \bar{A}$; hence $x \notin B$ (since $\bar{A} \cap B = \emptyset$). But then x must be in A, proving A closed in $A \cup B$. An identical argument proves B closed in $A \cup B$. Now assume A, B separated. If $x \in \bar{A} \cap B$, let $x_n \in A$, $x_n \to x$. Since $x \in B \subseteq A \cup B$ and A is closed in $A \cup B$, we must have $x \in A$, so $A \cap B \neq \emptyset$, a contradiction. Therefore $\bar{A} \cap B = \emptyset$, and similarly $A \cap \bar{B} = \emptyset$.

2. Assume $\bar{A} \cap B = A \cap \bar{B} = \emptyset$. If $G_1 = (\bar{B})^c$ and $G_2 = (\bar{A})^c$, then since $A \cap \bar{B} = \emptyset$ we have $A \subseteq (\bar{B})^c = G_1$, and since $\bar{A} \cap B = \emptyset$ we have $B \subseteq (\bar{A})^c = G_2$. Now $G_1 \cap B = (\bar{B})^c \cap B \subseteq (\bar{B})^c \cap \bar{B} = \emptyset$, and $G_2 \cap A = (\bar{A})^c \cap A \subseteq (\bar{A})^c \cap \bar{A} = \emptyset$. Now assume there exist open sets G_1, G_2 with the specified properties. Then (since $A \cap G_2 = \emptyset$) $A \subseteq G_2^c$ (closed); hence $\bar{A} \subseteq G_2^c$. Since $B \subseteq G_2$, we have $\bar{A} \cap B = \emptyset$. Similarly, $B \subseteq G_1^c$ (since $B \cap G_1 = \emptyset$), so $\bar{B} \subseteq G_1^c$. Since $A \subseteq G_1$, we have $A \cap \bar{B} = \emptyset$.

3. If $x \notin E$, then $E = [E \cap (-\infty, x)] \cup [E \cup (x, \infty)]$. Since $(-\infty, x)$ and (x, ∞) are open in R, $E \cap (-\infty, x)$ and $E \cap (x, \infty)$ are open in E (Section 8.1, Problem 1). Since E is connected, either $E \cap (-\infty, x) = \emptyset$ or $E \cap (x, \infty) = \emptyset$. In the first case, $y \in E$ implies $y \geq x$, so x is a lower bound of E, contradicting $a = \inf E$. In the second case, x is an upper bound of E, contradicting $b = \sup E$.

4. If $f(E) = A \cup B$, where A and B are disjoint and both open (and closed) in $f(E)$, then $E = f^{-1}(A) \cup f^{-1}(B)$, where $f^{-1}(A)$ and

$f^{-1}(B)$ are disjoint and (by continuity of f) both open (and closed) in E. Since E is connected, $f^{-1}(A) = \emptyset$ or $f^{-1}(B) = \emptyset$. But f maps onto $f(E)$, so either A or B is empty (e.g., if $y \in A, \exists x \in E$ with $f(x) = y$, so $x \in f^{-1}(A)$). Therefore $f(E)$ is connected.

5. Let f be a continuous, real-valued function on $[a, b]$, and assume $f(a) < c < f(b)$. By Problem 4, $f[a, b]$ is connected, so by Problem 3, $f[a, b]$ is an interval I. But $f(a), f(b) \in I$ implies $c \in I$, so $c = f(x)$ for some $x \in [a, b]$. Since $f(a) \neq c, f(b) \neq c$, we must have $x \in (a, b)$.

SECTION 8.4

1. Let $f_n(x) = 1 - n|x|, \ -1/n \leq x \leq 1/n$; $f_n(x) = 0$ elsewhere. Each f_n is continuous, but

$$\inf_n f_n(x) = \begin{cases} 0, & x \neq 0, \\ 1, & x = 0, \end{cases}$$

which is not LSC (it is USC).

2. Let $f_n(x) = nx, \ -1/n \leq x \leq 1/n$; $f_n(x) = -1, x \leq -1/n$; $f_n(x) = 1, x \geq 1/n$. Then

$$f(x) = \lim_{n\to\infty} f_n(x) = \begin{cases} 1, & x > 0, \\ 0, & x = 0, \\ -1, & x < 0, \end{cases}$$

and f is neither USC nor LSC.

3. Let $f(x) = 1/x, 0 < x \leq 1$; $f(0) = 0$.
4. f is LSC at x iff given $\epsilon > 0 \ \exists\delta > 0$ such that $d(x, y) < \delta \Rightarrow f(y) > f(x) - \epsilon$; f is USC at x iff given $\epsilon > 0 \ \exists\delta > 0$ such that $d(x, y) < \delta \Rightarrow f(y) < f(x) + \epsilon$. The proofs are done as in Theorem 4.1.2, with $d(f(x), f(y)) < \epsilon$ replaced by $f(y) > f(x) - \epsilon$ or $f(y) < f(x) + \epsilon$.
5. If f is nonnegative and finite–valued, then (see the proof of Theorem 8.4.6) so are the f_n. If $f \geq 0$ but can assume the value $+\infty$, map the nonnegative extended reals one-to-one onto $[0, \pi/2]$ via $h(x) = \arctan x$ and proceed exactly as in Theorem 8.4.6. The functions f_n will automatically be nonnegative.

REVIEW PROBLEMS FOR CHAPTER 8

1. (a) Always
2. (b) Sometimes (see Section 8.1)
3. (a) $\{1, \frac{1}{2}, \frac{1}{3}, \frac{1}{4}, \ldots\}$
   (b) Cantor set

INDEX

D

E

F

G

H

I

O

P

Q

R

S

T

U

V

W